SCIENCE SKILLS
Problems in GCSE Chemistry

D1824132

J Bennetts

Senior Lecturer in
Science Education
St Martin's College
Lancaster

M Hannon

Deputy Director of
CATS Key Stage 3 SAT
Development Team (Science)

J Mundie

Lecturer in Science and Engineering
East Warwickshire College
Rugby

Hodder & Stoughton

LONDON SYDNEY AUCKLAND TORONTO

Contents

Skills

Note

The following symbols appear beside some questions in the text. See Introduction for details:

✱ Questions which require some knowledge of facts, concepts and principles.

A Questions for which a discussion of the answer is provided.

Prefixes used in this book

Symbol	Prefix	Meaning	Example	
μ	micro	$\frac{1}{1\,000\,000}$	μm	(micrometre)
m	milli	$\frac{1}{1000}$	mm	(millimetre)
c	centi	$\frac{1}{100}$	cm	(centimetre)
d	deci	$\frac{1}{10}$	dm	(decimetre)
k	kilo	1000	km	(kilometre)
M	mega	1 000 000	MJ	(megajoule)

Contexts

The following table shows the context in which each question is set.

Context	Chapter 1	Chapter 2	Chapter 3	Chapter 4	Chapter 5
Acids, bases, salts	Q2 Q8	Q1 Q2 Q3 Q5 Q16 Q23	–	Q9	–
Agriculture & industry	Q22 Q23 Q26	Q7 Q15 Q19 Q24	Q1 Q2 Q6 Q6	Q15 Q16 Q17	Q1 Q4 Q13
Atmosphere & pollution	Q6 Q11	Q9 Q18 Q20	Q6 Q13 Q19 Q20	Q2 Q12 Q18	–
Chemical reactions	Q20	Q10 Q14 Q26	–	–	Q3 Q7 Q9 Q16 Q17
Metals	Q10	Q21	Q12	–	Q6 Q18
Mixtures, solutions & separation	Q1 Q9 Q12 Q14 Q24 Q27	Q12 Q13 Q22 Q25	Q9	Q2 Q3 Q7 Q14	Q5 Q7 Q10 Q11 Q14 Q17
Periodic table & atomic structure	Q29	Q4 Q6	Q3 Q4 Q7 Q15 Q16	–	–
Properties of materials	Q3 Q5 Q7	–	Q5 Q8	Q1 Q5 Q8 Q10 Q11 Q18	Q8
Quantitative chemistry	Q15 Q28	Q8 Q21 Q23	–	–	–
Rates of reaction	Q25	Q17	–	–	–
Resources, fuels & energy	Q13 Q16 Q17 Q19 Q21	Q11	Q10 Q13 Q14	Q4 Q15	Q15

Introduction: mainly for teachers

The *Science Skills* series is designed to cultivate the ability to handle information, solve problems and perform the other processes which are essential to science disciplines. National Criteria for GCSE Chemistry and other sciences require that no more than 25% of the assessment be based on the recall of facts, theories, principles and concepts. Coursework is weighted at between 20% and 30%. The books in this series seek to develop and exercise all those attributes which make up the remaining 50% of GCSE assessment.

We feel that it is important for students preparing for GCSE to experience science in real contexts. Whenever possible, the questions in this book put Chemistry in a real context. National Criteria require that technological applications and social, economic and environmental issues pervade GCSE assessment in science subjects.

The rationale for the grouping of questions under our five chapter headings is given in the Introduction to *Science Skills: Problems in GCSE Science*. It is intended as a way of blocking together the objectives of National Criteria and relates to the epistemology of many syllabuses. However, we freely admit that the placement of a question under a particular heading is somewhat approximate. Science does not operate in discrete packages. It is not easy to decide whether analysis of a set of data to find a rogue result is information handling or finding patterns.

The authors of the *Science Skills* series are committed to the aims of Broad Balanced Science. *Problems in GCSE Science, Biology, Chemistry* (and soon *Physics*) attempt to give a coherence to the skills which are common to all the sciences. At present most examinations for The Sciences Double Award contain some integrated questions. (This includes LEAG 'Combined' and MEG 'Combined', both of which have separate Biology, Chemistry and Physics syllabus sections.) However, the majority of questions in these examinations are identifiable as relating to 'Chemistry' or one of the other subjects. If all questions on the Periodic Table had to involve some Biology and Physics, some of them would be very contrived. This book therefore looks at Chemistry questions in the context of Balanced Science, in which interplay of common skills is fundamental.

With a few questions in this book the content naturally links to Biology or Physics and this development has been permitted. For example, if a student is to make sense of the Chemical aspects of ozone depletion in the atmosphere, it is reasonable that the Physics reasons for concern about this effect are considered. This indeed is the essence of GCSE Chemistry which takes account of technological, social, economic and environmental factors. Therefore, the book will be of value to those following a GCSE Chemistry course irrespective of the Balanced Science issue.

Looking ahead to Science in the National Curriculum, the general introduction to the programme of study for key stage 4 states:

> The abilities to communicate, to apply, to investigate and to use scientific and technological knowledge and ideas to make informed judgement are essential elements of the study of science.

It is the purpose of this book to provide opportunities for the development of these abilities.

The first four chapters begin with multiple choice questions. These are arranged in an incline of difficulty. The questions which follow are also arranged in order of increasing difficulty. But it should be noted that there is a discontinuity. The multiple choice questions get progressively harder. The earlier non-multiple choice questions are then very easy.

Most of our questions are 'recall free'. Those which require some knowledge of facts, concepts and principles are marked with an asterisk ✳ .

[A] A number of questions are flagged with this letter. In the final chapter the answering of these questions is discussed. In most cases, these are not simply mark schemes, but rather an analysis of what the question was designed to test and advice on how to approach it.

The authors have each taught Integrated Science and two have taught Chemistry to A-level. Two have been Heads of Science in Comprehensive Schools. The three authors are each currently Chief Examiners for Double Award GCSE Balanced Science syllabuses.

1 Information handling

1

Which chemical is there most of in a packet of 'Dad's Washing Powder'?

A. Alkylbenzenesulphonate.
B. Sodium silicate.
C. Sodium sulphate.
D. Sodium tripolyphosphate.

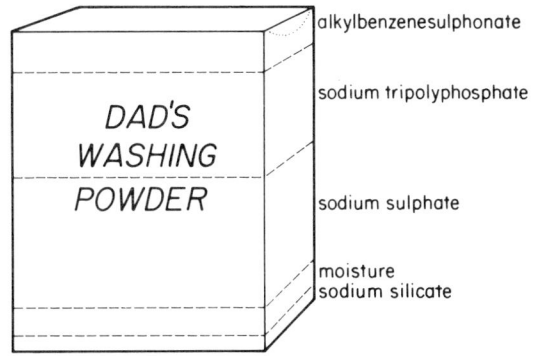

2

The pictogram below gives information about the use of sulphuric acid by different industries in Britain per year.

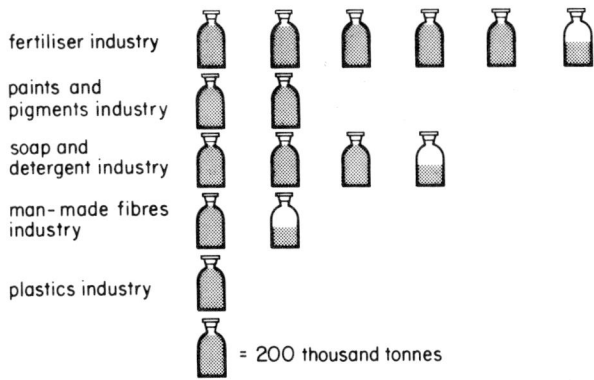

= 200 thousand tonnes

(a) The greatest use of sulphuric acid is for making:
   A.   fertilisers.

B.   man-made fibres and plastics.
C.   soaps and detergents.
D.   paints and pigments.

(b) How much sulphuric acid is used each year in Britain for making man-made fibres?

A.   1½ bottles.
B.   2 bottles.
C.   300 thousand tonnes.
D.   400 thousand tonnes.

3

Mr Patel buys materials for a company which makes bottles for transporting chemicals, including acids and alkalis. He sees information about a new material called 'Glassthene'. Some of the properties of 'Glassthene' are given below.
Which one would make Mr Patel decide not to buy 'Glassthene'?
A. It is not biodegradable.
B. It is not clear.
C. It is attacked by acid in sunlight.
D. It can withstand rough handling.

4

Element	Colour of its compounds	Colour it gives flames
Calcium	no colour	red
Copper	blue/green	green
Sodium	no colour	yellow
Zinc	no colour	none

From the information in the table above, you can say that

A. only colourless compounds colour a flame.
B. only coloured compounds colour a flame.
C. colourless compounds can colour a flame.
D. flame colour is the same as the colour of the compound.

1

5

The table below gives some information about four materials. Use it to help you answer the questions.

Material	Conducts heat	Conducts electricity	Transparent	Easy to bend and shape
A	×	×	√	×
B	√	√	×	√
C	√	×	√	×
D	×	×	×	√

√ = yes, × = no.

(a) Which material could be used to make a saucepan? *(1)*

Explain your answer. *(2)*

(b) Electric cables contain a core material which carries the current. This is surrounded by a covering material which stops you from getting an electric shock.

Which material would you choose for making

(i) the covering? *(1)*

(ii) the core? *(1)*

(c) Which material would be best for making house windows in a cold area? *(1)*

Explain why. *(2)*

6

(a) Diagram 1 gives information about some of the man-made air pollution produced each year in Britain.

(i) The gases that produce acid rain come mainly from:

A. vehicles.
B. industry.
C. power stations.
D. volcanoes.

(ii) The amount of nitrogen oxides, in million tonnes, produced by vehicles is:

A. 0.51
B. 1.07
C. 2.64
D. 7.10

Diagram 1

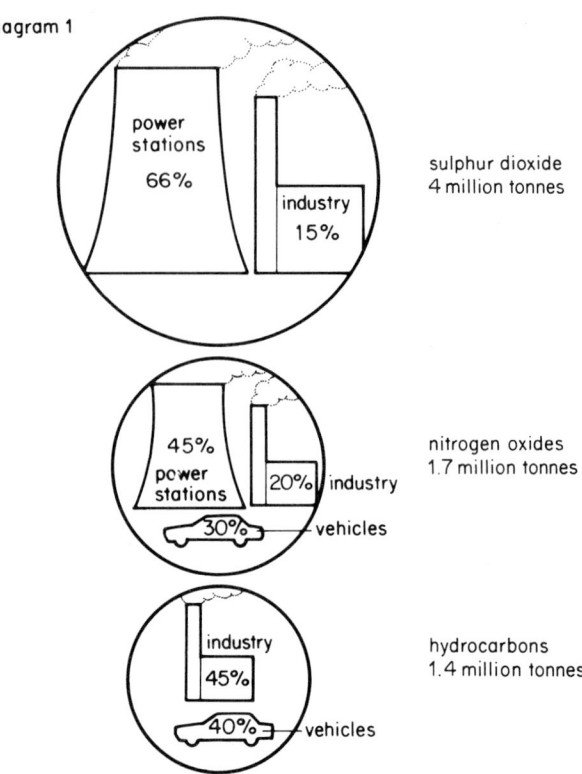

sulphur dioxide
4 million tonnes

nitrogen oxides
1.7 million tonnes

hydrocarbons
1.4 million tonnes

(b) Diagram 2 shows estimates of the amount of these gases produced each year on Earth. These gases are also produced by nature.

Diagram 2

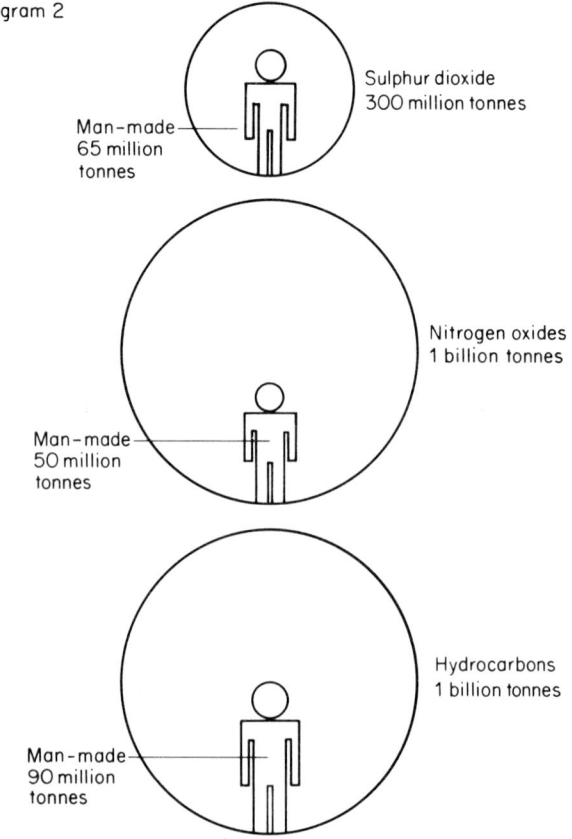

Sulphur dioxide
300 million tonnes

Man–made
65 million
tonnes

Nitrogen oxides
1 billion tonnes

Man–made
50 million
tonnes

Hydrocarbons
1 billion tonnes

Man–made
90 million
tonnes

Which of these statements is correct?

  A. Sulphur dioxide is mostly man-made.
  B. The amounts of nitrogen oxides and hydrocarbons produced by nature are the same.
  C. Less than 10% of these gases are man-made.
  D. The amount of nitrogen oxides produced by nature is more than four times the amount of sulphur dioxide produced by nature.

7

The table below gives some data about five substances A–E.

Substance	Melting point (°C)	Boiling point (°C)	Electrical conduction		
			solid	liquid (melted)	solution in water
A	1455	2837	good	good	does not dissolve
B	−56	−28	poor	poor	poor
C	685	1324	poor	good	good
D	−51	−35	poor	poor	good
E	−89	118	poor	poor	poor

Which of these substances fit the following descriptions? There may be more than one substance fitting each description.
(a) (i) A solid at room temperature.
    (ii) A liquid at room temperature.
    (iii) A gas at 150 °C.   (6)
(b) (i) A metal.
    (ii) Covalent compounds.
    (iii) Ionic compounds.   (5)

8

Soil can contain nitrogen, phosphorus, potassium and other elements. But if the pH of the soil is too high or too low, plants are not able to take up these elements.

The charts below show how the amount of an element which can be taken up by plants depends on the pH of the soil. The narrower the bar, the harder it is for plants to take up the element.

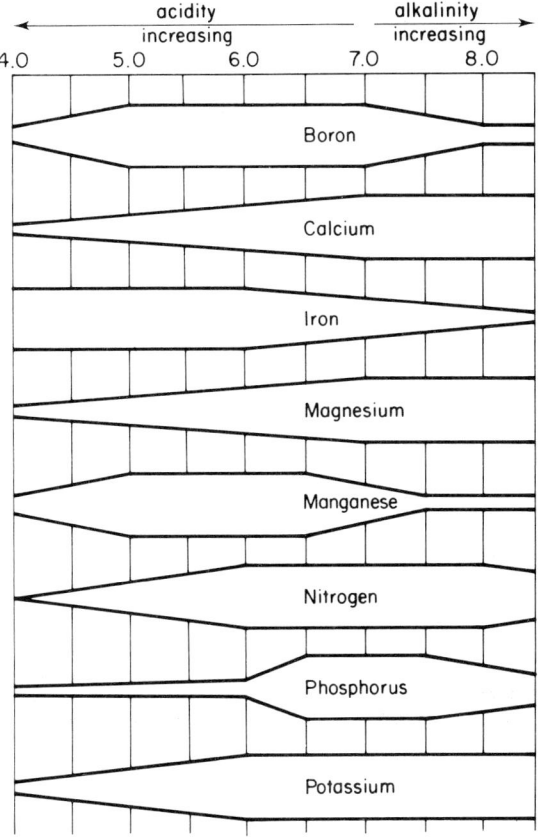

(a) Which elements cannot be taken up easily if the soil pH is 8?
  A. Calcium, magnesium, nitrogen, phosphorus.
  B. Boron, iron, magnesium, phosphorus.
  C. Boron, iron, manganese, phosphorus.
  D. Boron, iron, manganese, nitrogen, phosphorus.

(b) Which elements cannot be taken up easily if the soil pH is 6?
  A. Boron, iron, manganese, nitrogen, potassium.
  B. Calcium, magnesium, phosphorus.
  C. Calcium, magnesium, phosphorus, potassium, nitrogen.
  D. Calcium, manganese, phosphorus.

(c) Most rhododendron plants cannot be grown at a pH higher than 6. This could be because they need a lot of:
  A. iron.
  B. nitrogen.
  C. phosphorus.
  D. potassium.

A *9

Bronzes are alloys of copper, tin and often other metals. The table below gives data about the relative amounts of each element in different alloys.

Alloy	Copper Cu	Tin Sn	Zinc Zn
Bronze (Admiralty gunmetal)	44	5	1
Bell bronze (used in making church bells)	5	1	0
Speculum metal (used in making metal mirrors)	2	1	0
Coinage bronze (used for making 1p and 2p pieces)	95	4	1

Alloys are mixtures. But the compounds Cu_3Sn and Cu_4Sn exist.
(Relative atomic masses: $Cu = 64$; $Sn = 118$)
Which of the alloys in the table is most likely to contain both Cu_3Sn and Cu_4Sn?

A. Admiralty gunmetal.
B. Bell bronze.
C. Speculum metal.
D. Coinage bronze.

*10

In an experiment, pupils used samples of three different metals, P, Q and R. They made three cells as the diagram shows.

voltmeter readings

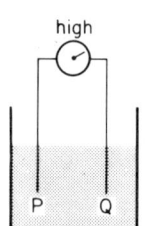 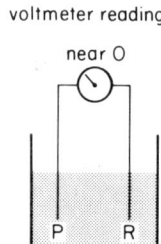 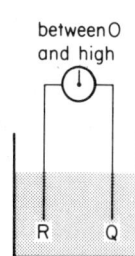

high   near O   between O and high

beakers of electrolyte : sodium sulphate solution

(a) From the readings of the voltmeters, the order of reactivity of the metals is:

A. PQR.
B. PRQ.
C. QPR.
D. RPQ.

(b) The metals are in the electrochemical series:
magnesium (2.37 volt), aluminium (1.66 volt), zinc (0.76 volt), iron (0.44 volt), nickel (0.25 volt), tin (0.14 volt), copper (–0.15 volt), silver (–0.8 volt).
Choose from the table below which is most likely to be P, Q and R.

	P	Q	R
A.	iron	aluminium	zinc
B.	copper	zinc	iron
C.	aluminium	zinc	iron
D.	zinc	iron	copper

11

In England and Wales cancer rates are higher in towns than in the countryside. Some scientists think that this causes ten thousand deaths a year in towns.

This is more than the total number of cancer deaths anticipated in the USSR over the next sixty years as a result of the nuclear accident at Chernobyl.

In Glasgow, the death rates from the seven most common cancers are 40% above those in the Scottish Western Isles where people breathe clean air. Some scientists think that these cancers are caused by complex nitrogen compounds in vehicle exhaust fumes and from badly-adjusted fossil fuel burners in homes and factories.
(Based on an article by Professor John Fremlin in *Physics Bulletin*, Vol. 39, No. 4. April 1988)

(a) Which of these statements is correct?

A. 10 000 people die each year in the towns of England and Wales.
B. 10 000 more people die in the towns of England and Wales than in the countryside.
C. 10 000 people die of cancer in the towns of England and Wales.
D. If cancer rates were the same in towns as in the countryside, 10 000 fewer people would die each year in England and Wales.

(b) The Chernobyl accident happened in April 1986.

A. It killed 10 000 people in 1986.
B. It may have killed nearly 10 000 people by the year 2046.
C. It may have killed nearly 600 000 people by the year 2046.
D. It may kill more people than those dying in towns in England and Wales in a year.

(c) Which of these statements is correct?

    A. In Glasgow the death rate is 40% higher than in the Western Isles.
    B. In Glasgow 40% of the people get cancer.
    C. In the Western Isles people do not get cancer.
    D. In the Western Isles the death rate from the seven most common cancers is only 71.4% as high as in Glasgow.

(d) Which of these statements is NOT correct?

    A. The cancer rate in towns could be reduced if people had their central heating boilers serviced more frequently.
    B. The cancer rate could be reduced if towns did a 'park and ride' system so that cars had to be parked on the outskirts and buses used to get to the town centre.
    C. The cancer rate would be lower if each town still had its own coal-fired power station instead of the big power stations in the National Grid.
    D. The cancer rate in towns could be reduced if more vehicles had battery-operated electric motors rather than petrol engines.

*12

The pie chart below shows the ways in which copper is used.

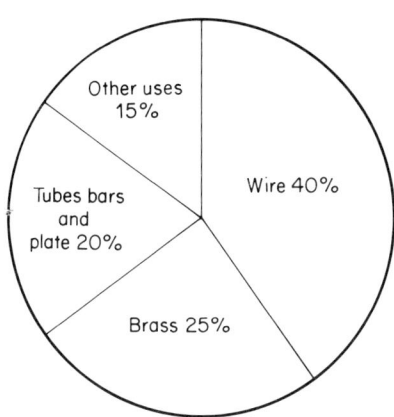

(a) (i) What % of copper is used to make brass? *(1)*
   (ii) Which other element is used to make brass? *(1)*
(b) (i) Which is the most important use of copper? *(1)*
   (ii) Why is so much copper used for this purpose? *(1)*

13

The data below show the percentages of energy from different sources used in the United Kingdom during 1985.

Source	Percentage
Coal	32.2
Oil	35.2
Natural gas	25.2
Nuclear & hydroelectric	7.4

(a) (i) Which is the most important source of energy?
   (ii) What percentage of energy comes from natural gas? *(2)*
(b) Draw a pie chart for this data. *(4)*
(c) (i) Which one of the sources of energy in the table is not used to produce electricity?
   (ii) Which energy source is renewable? *(2)*

14

Diagram 1 below shows the percentage of materials used to make glass.

Diagram 1

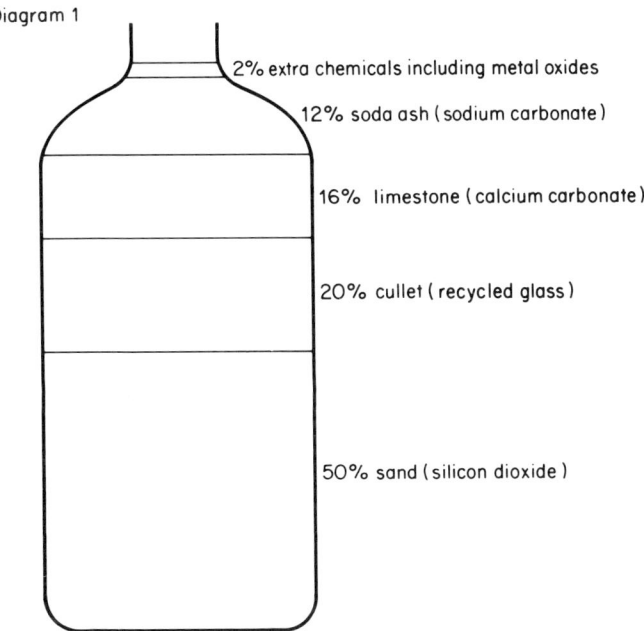

(a) Plot a bar chart to show the information given in the diagram. *(4)*

(b) Diagram 2 below shows a bottle bank. These are often seen on car parks where they have been provided by local councils.

Diagram 2

green   clear   brown

B O T T L E   B A N K

(i) Explain the purpose of a bottle bank. Refer to the data given in diagram 1. (2)

(ii) Why are there three separate holes labelled green, clear and brown? (1)

*15

Use the information in the table to answer this question about fertilisers.

Fertiliser	How quickly it dissolves in water
KNO_3	dissolves quickly
$(NH_4)_3PO_4$	dissolves quickly
$CO(NH_2)_2$	dissolves slowly
NH_4NO_3	dissolves quickly

(a) Copy this table and complete it by matching the chemical name with the correct formula.

Chemical name	Formula
Ammonium nitrate	
Ammonium phosphate	
Potassium nitrate	
Urea	

(4)

(b) Which of these fertilisers contains the most nitrogen?
(Relative atomic masses: $H = 1$; $C = 12$; $N = 14$; $O = 16$; $K = 39$)
Show how you arrived at your answer. (5)

(c) Which of the fertilisers would be best on sandy soil? Explain your answer. (2)

*16

The bar graphs below show the percentage of different gases in four natural gas fields.

(a) Which field contains the largest percentage of methane?
(b) Which field's gas contains no propane?
(c) What percentage of nitrogen does the gas found in Holland contain?
(d) Explain how the gas found in France could produce sulphur dioxide (SO_2).
(e) Suggest why the energy given out by gas from Holland is less than from North Sea gas.

17

(a) Diagram 1 shows for four countries the percentages (%) of the total energy need which is met by burning wood.

Diagram 1

(i) Which country gets about half its energy from wood?
   A. India.
   B. Kenya.
   C. Sweden.
   D. USA. *(1)*

(ii) Which country gets about a quarter of its energy from things other than wood?
   A. India.
   B. Kenya.
   C. Sweden.
   D. USA. *(1)*

(b) Diagram 2 shows for four countries the percentage of electrical energy which comes from nuclear power stations.

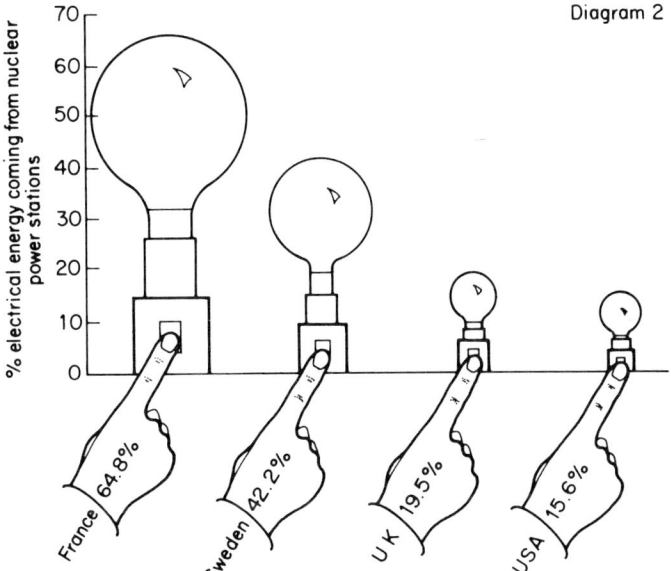

Diagram 2

Which of these statements is based on correct use of this data?

A. The total amount of nuclear-generated electricity produced in the USA is less than that produced in the UK.
B. The total amount of nuclear-generated electricity produced in Sweden is more than that produced in the UK.
C. The UK will never get most of its power from nuclear power stations.
D. France gets more than a third of its electricity from non-nuclear sources. *(1)*

(c) Use the information in the diagrams to explain these statements.
   (i) In France, electricity is 30% cheaper than in the UK. *(2)*
   (ii) The USA depends heavily on fossil fuels. *(2)*
   (iii) At least 9% of Sweden's energy comes from renewable resources. *(2)*

18

This table shows the amount of water used by an average person each day.

Use	Amount used per person (dm³)
Drinking and cooking	5
Flushing the lavatory	50
Washing clothes	15
Washing and bathing	50
Washing the dishes	15
Watering the garden and washing the car	10

(a) Plot a bar chart using the information above. *(8)*
(b) What is the total volume of water that a person uses in one day? *(2)*
(c) If there has been no rain, water supplies get low. Using a hosepipe to wash cars and water the garden is usually banned.
   (i) Explain how this ban should help. *(1)*
   (ii) Suggest and explain ONE other way in which people could save water. *(2)*
(d) Some people use more water than others, and the amount of water that they use depends on the time of year.
   (i) In what occupations are people likely to use a lot of water? *(1)*
   (ii) At what time of year is most water likely to be used? *(1)*

19

Crude oil has many important uses. The amounts used for different purposes are given below as percentages.

Heating homes	5%
Heating and power for industry	20%
Heating and power for other consumers	20%
Making chemicals	10%
Making electricity	10%
Road transport	35%

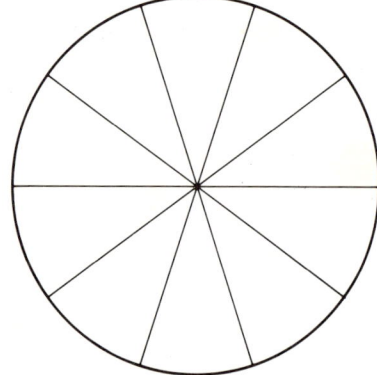

Copy out the pie chart. Each segment is worth 10%.
(a) Use the data above to complete the pie chart. Shade and label it or give a key. (6)
(b) Scientists are trying to find alternative sources of energy to oil. Suggest two reasons why this work is important. (2)

*20

All green plants carry out a reaction called photosynthesis. To do this they need to contain a substance called chlorophyll and to have light energy. The equation for photosynthesis is

$$6CO_2 + 6H_2O + \text{light energy} \rightarrow C_6H_{12}O_6 + 6O_2$$
$$\text{glucose}$$

Like all living things, plants also carry out respiration. The equation for this reaction is

$$C_6H_{12}O_6 + 6O_2 \rightarrow 6CO_2 + 6H_2O + \text{energy}$$

Scientists investigated these reactions using carbon dioxide made from ^{14}C.
(a) (i) What is ^{14}C? (1)
    (ii) Suggest how this will make the carbon dioxide used in the investigation different from ordinary carbon dioxide in air. (2)

(b) As part of the investigation, a plant was left in a clear glass container full of this carbon dioxide.
    Explain carefully what you would expect to happen to the amount of carbon dioxide in the container if it was:
    (i) kept in the light. (3)
    (ii) kept in the dark. (3)
(c) (i) Explain what is meant by the word:
    exothermic (2)
    endothermic (2)
    (ii) Match the words in c(i) to the plant's reactions:
    Photosynthesis is . . . .
    Respiration is   . . . . (1)

*21

This bar chart shows the percentage of various elements in the Earth's crust.

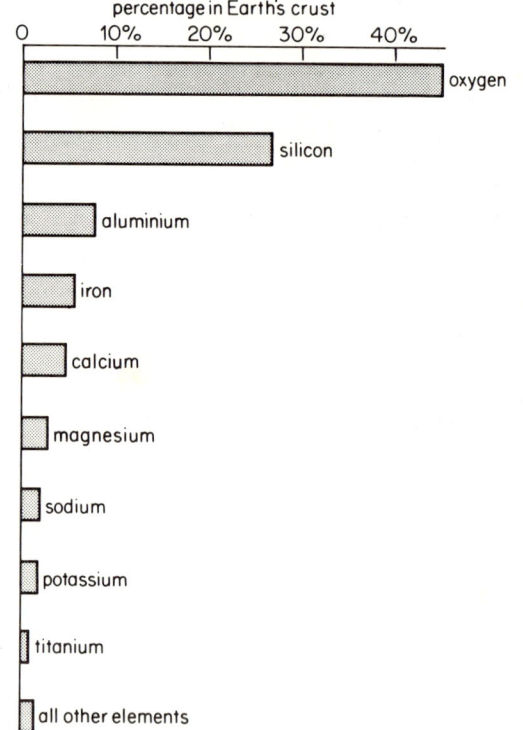

(a) Which is the most common element? (1)
(b) Which is the most common metal? (1)
(c) What percentage of the Earth's crust is calcium? (1)
(d) The Earth's crust contains more aluminium than iron, yet iron is much cheaper. Suggest and explain two reasons for this. (2)

*22 Tanning leather

Hair is made from a protein called keratin. The outer layer of skin, called the epidermis, is also made of keratin. Lower layers of skin are made of another protein called collagen. Fairly weak alkalis, such as calcium hydroxide, dissolve keratin but not collagen. In making leather, a solution with a pH of about 12 is used. This removes hair from the animal skin.

Washing can remove the alkali from the skin. But when calcium hydrogencarbonate in hard water comes into contact with calcium hydroxide, calcium carbonate is formed. This forms a white deposit on the skin and spoils it. So leather making factories need soft water.

A skin is hard and stiff when it is dry but leather is flexible. The tanning process turns the skin into leather. Proteins in skin are very long molecules. They are twisted together and cross over one another. If there is moisture in the skin, water molecules stop the protein molecules joining up so the skin is flexible. When the skin is dried, bonds form between protein molecules and the skin becomes stiff. If certain metal ions such as chromium and aluminium are absorbed by the skin, they come between the protein molecules. This stops the protein molecules joining up. So the skin stays soft when it dries. Substances which turn skin into leather are called tans. Chromium (III) sulphate and aluminium potassium sulphate are tans.

Some materials from plants are also tans. They contain substances called tannins. In Europe, suitable materials can be obtained from the wood of sweet chestnut trees and from the bark of pine trees, but the bark of oak trees is best. 'Oak apples' are small brown balls found hanging from some oak trees. They are caused by gall wasps and they too contain a lot of tannin. Wattle trees grow in Australia and many parts of Africa. Their bark contains tannins. Dried Indian plums called myrobalans also contain tannins.

(a) Calcium hydroxide removes hair from a skin. What else do you think it removes? *(1)*
(b) Suggest why alkalis feel slippery. *(2)*
(c) Write a complete word equation for the reaction between calcium hydroxide and calcium hydrogencarbonate. *(2)*
(d) There are more leather tanneries in some areas than in others. Suggest two possible reasons for this. *(2)*
(e) Washing with boric acid solution lowers the pH faster than washing with water. Why is this? *(2)*
(f) Dried skins are stiff because protein molecules in the skin have joined together.
　(i) What is the name of this protein? *(1)*
　(ii) Before the protein starts to join up it is a monomer. What do we call the material formed when many similar molecules join to make a big molecule? *(1)*
(g) Name a metal ion which can act as a tan. *(1)*
(h) (i) Two thousand years ago, each Roman town in Britain had a tannery. Suggest a material which the Romans could have used to tan leather. *(1)*
　(ii) At the same time, people in the mountainous regions of north India were tanning leather. What might they have used for this purpose? *(1)*
(i) Oak apples are made by wasps. Why do you think oak apples contain tannin? *(1)*

23

The figures in the table below refer to crops grown in the United States of America in one year and the cost of pesticide chemicals used to treat them (in dollars).

| | CROP | | | |
	Cotton	Soya bean	Corn	Wheat
Acres harvested (in millions)	10	71	73	79
Cost (in millions of dollars) of				
Herbicides	80	850	790	110
Insecticides	162	30	190	10
Fungicides	3	10	1	8
Other	53	20	19	2
Total pesticide cost (in millions of dollars)	298	910	1000	130

(a) Draw a bar chart to show these figures. *(4)*
(b) Explain the meaning of the terms
　(i) pesticide
　(ii) herbicide
　(iii) insecticide
　(iv) fungicide. *(4)*
(c) A friend says that the 'other' row in the table must represent the cost of fertilisers used. Explain whether this conclusion could be correct. *(2)*
(d) (i) For each crop, calculate the cost of pesticides per acre harvested. Arrange these costs in order with the most expensive first. *(5)*
　(ii) Explain two other factors which would need to be taken into account before you could decide which one of these crops is the most expensive to grow. *(4)*

24

The graph below shows the maximum mass of potassium chloride that can be dissolved in 100 g of water at different temperatures.

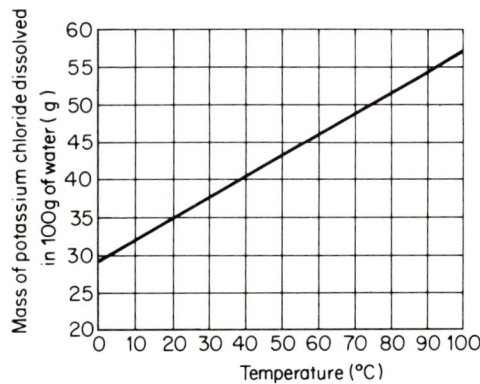

(a) How much potassium chloride dissolves in 100 g of water at 60 °C? (1)
(b) At what temperature will 40 g of potassium chloride just dissolve in 100 g of water? (1)
(c) How much water at 90 °C will just dissolve 150 g of potassium chloride? (2)
(d) What would you notice when a solution of 50 g of potassium chloride in 100 g of water is cooled from 80 °C to 20 °C? (3)

*25

A group of pupils were investigating what happened when hydrogen peroxide and manganese (IV) oxide were mixed together. In one experiment they measured the volume of gas produced when a small sample of manganese (IV) oxide was added to some hydrogen peroxide. They made a note of the volume of gas every 20 seconds and put the results into this table.

Time (seconds)	0	20	40	60	80	100
Volume of gas (cm³)	0	5	15	25	35	35

(a) Draw a graph to show these results. (5)
(b) (i) What is the total volume of gas produced in this reaction? (1)
 (ii) How long did the reaction take to finish? (1)

(c) The pupils decide to do two more experiments:
Experiment M, doing the experiment at the same temperature but using a greater volume of hydrogen peroxide.
Experiment T, using the same amount of each chemical but doing the experiment at a lower temperature.
(i) Add to your graph a sketch graph, labelled M, to show the results that you would expect for Experiment M. (3)
(ii) Add another sketch graph, labelled T, to show the results you would expect for Experiment T. (4)

26

Plants need the minerals nitrogen, phosphorus and potassium. These minerals must be available to crops for farm land to produce a good yield. For grass and wheat, nitrogen is the most important mineral.
(a) The cyclic polygons in diagram 1 show the number of kilograms of nitrogen taken from a hectare of land per day when grass or wheat are growing really well.

Diagram 1(a)

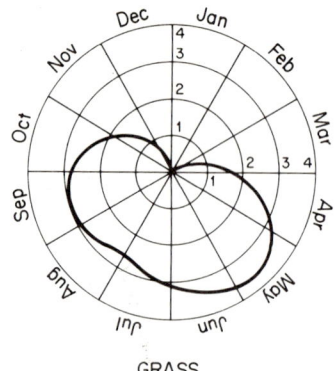

GRASS

Diagram 1(b)

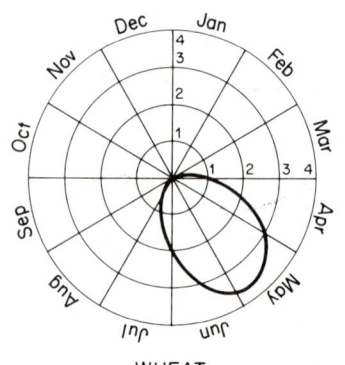

WHEAT

(i) At what time of year is the most nitrogen
    needed by
    1. grass? *(1)*
    2. wheat? *(1)*
(ii) At what time of year is nitrogen first needed by
    1. grass? *(1)*
    2. wheat? *(1)*
(iii) At what time of year is nitrogen no longer
    needed by
    1. grass. *(1)*
    2. wheat? *(1)*
(iv) Suggest explanations for the shape of the
    graphs for
    1. grass. *(3)*
    2. wheat. *(2)*

(b) Soil contains about 3000 kg of nitrogen per hectare
in the top 15 cm. About 98% of this is not available
to plants because it is organic nitrogen which is
bound to certain carbon atoms in humus.
    Bacteria convert this organic nitrogen to nitrate
ions which are then available to plants. This
process is called 'mineralisation'. The activity of
these bacteria depends on the temperature,
moisture and air conditions in the soil.
(i) Estimate the amount of nitrogen available to
    plants in the top 15 cm of a hectare of soil. *(2)*
(ii) Farmers plough their fields. This digs and
    turns the soil over. Ploughing increases
    mineralisation. Suggest why. *(2)*
(iii) The cyclic graph in diagram 2 shows the
    number of kilograms of nitrogen produced
    per hectare per day by mineralisation.

Diagram 2

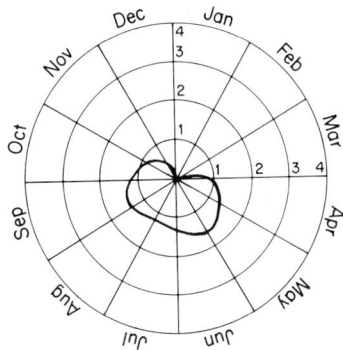

1. When is mineralisation fastest? *(1)*
2. Describe how the rate of mineralisation
   changes during the year. *(4)*
3. Small quantities of nitrate enter the soil in
   rain. Clover and some other crops have
   organisms in their roots which can convert
   nitrogen from the air into nitrates in the
   soil. This is called 'fixing' nitrogen. Clover

can fix 250 kg of nitrogen per hectare in a
year. Discuss whether the information
given suggests that artificial fertilisers are
needed to get the best out of farm land. *(6)*
4. Compare diagrams 1 and 2 to suggest why
   grass is the most common plant in the
   countryside. *(4)*

27

Table 1 shows the solubility of different substances at
different temperatures.

Table 1

Substance	Solubility * at				
	0 °C	20 °C	40 °C	60 °C	80 °C
Copper sulphate	14	21	29	40	55
Potassium nitrate	13	32	64	110	169
Sodium chloride	36	36	36	37	38
Oxygen from the air	0.0014	0.00089	0.00065	0.00056	–

* Solubility measured as the number of grams of substance which
dissolves in 100 g of water.

(a) Which substance is the most soluble at
    (i) 20 °C? *(1)*
    (ii) 60 °C? *(1)*
(b) Which substance's solubility is most affected by
    temperature? *(1)*
(c) Estimate the solubility of oxygen at 80 °C. *(2)*
(d) Sea water is 'salty' because it has sodium chloride
    dissolved in it. It also contains oxygen, which is
    important to fish. The temperature of sea water in
    the English Channel is about 10 °C. The
    temperature of sea water in the Bay of Bengal is
    about 35 °C. Table 2 gives data about the mass of
    sodium chloride and oxygen dissolved in 100 g of
    sea water in these two areas.

Table 2

Place	Sodium chloride (g)	Oxygen (g)
English Channel	2.72	0.0009
Bay of Bengal	2.80	0.00042

(i) Use the information in table 1 to explain the
    difference in sodium chloride and oxygen
    levels in these two areas. *(5)*
(ii) If a solution cannot dissolve any more of a
    substance, we say that it is saturated. How
    saturated with sodium chloride and oxygen is
    the sea water in these two areas? *(2)*

28

In an investigation of the reaction between copper and oxygen, some pupils got these results:

Mass of copper used (g)	Mass of copper oxide made (g)	Mass of oxygen used (g)
3.2	4.0	–
6.4	8.0	–
12.8	16.0	–

(a) Copy and complete the table to show the mass of oxygen used in each experiment. (2)

(b) Copper can form four different oxides. Which was formed in this investigation? (Relative atomic masses: O = 16; Cu = 64) (2)

  A. Cu_2O
  B. CuO
  C. CuO_2
  D. Cu_2O_3

(c) Read this passage and answer the questions below:

  Copper was being used for making tools and weapons 4000 years ago. It was usually mixed with tin to form bronze. Bronze melts at a lower temperature than copper, so it is more easily made into tools and other things.

  The Greeks and Romans obtained copper from the island of Cyprus. The Romans called it cyprium and later cuprum. The Roman goddess of love, Venus, was associated with Cyprus. Arab chemists in the early Middle Ages (alchemists) called copper 'venus' and they gave it the symbol ♀ which they also used to represent a mirror. Today this symbol is used by biologists to mean 'female'. These Arab alchemists made many important scientific discoveries. In particular, they left records about their experiments on copper. For example: 'We can polish ♀ metal to give a very shiny surface. But when it is heated in a hot fire, ♀ forms black scales which are red on the inside. The material is heavier after heating than before. A piece of ♀ with a mass of 100 drachm is heated. The scales are scraped off. They have a mass of 1.9 drachm and the ♀ left has a mass of 98.4 drachm.'

  (i) What is bronze made from? (1)
  (ii) Explain why bronze is better than copper for making tools. (2)
  (iii) Copper compounds used to be called cuprous or cupric. How did they get this name? (1)
  (iv) Suggest why the alchemists used the same symbol for copper and for a mirror. (1)
  (v) The alchemists did not measure mass in kilograms. What unit did they use? (1)
  (vi) Compare the data left by the alchemists with that in part (b) of this question. Which oxides of copper were formed in the alchemists' experiments? Explain your answer. (4)

*29

Diagram 1 shows the first three periods of the modern Periodic Table.

Diagram 1

Group / Period	I	II				III	IV	V	VI	VII	O
1				H 1							He 2
2	Li 3	Be 4				B 5	C 6	N 7	O 8	F 9	Ne 10
3	Na 11	Mg 12				Al 13	Si 14	P 15	S 16	Cl 17	Ar 18

The Periodic Table used to be shown as in diagram 2.

Diagram 2

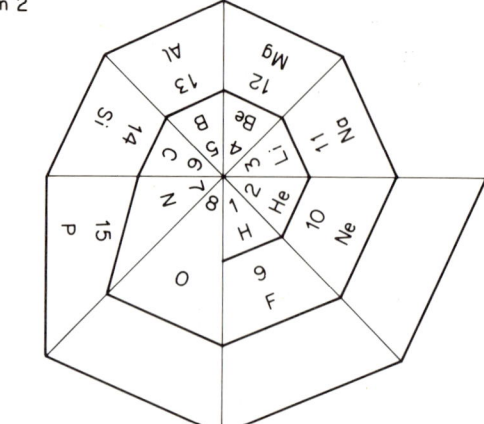

(a) Copy and complete diagram 2 to show the elements up to argon. (3)

(b) Label the sections of diagram 2 to show the different groups in the Periodic Table. (1)

(c) In what way is the position of hydrogen unsatisfactory? (1)

(d) Period 4 starts with potassium (atomic number 19). Can potassium be shown on diagram 2 in a satisfactory way? (1)

2 Problem solving

*1

Local environmental health regulations require that waste water put into the public drains should have a pH between 8 and 9. A check on a factory's waste water showed the pH to be 13.5. Which of the following should be added to change the pH of the waste water to the required level?
A. Calcium carbonate.
B. Hydrochloric acid.
C. Potassium nitrate.
D. Sodium chloride.

*2

Burning oil or coal produces sulphur dioxide, which can be removed by bubbling the waste gas through a liquid.

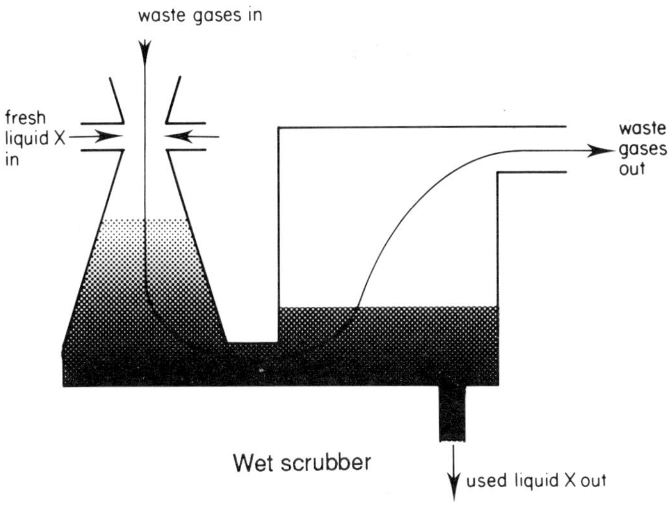

Wet scrubber

To remove sulphur dioxide, X should be
A. sodium carbonate solution.
B. sodium chloride solution.
C. sodium nitrate solution.
D. sodium sulphate solution.

*3

Maggie Jones has spilt some ammonia solution in a cupboard in the kitchen. Which of the following common household substances could she use to neutralise the ammonia?
A. Common salt (sodium chloride).
B. Washing soda (sodium carbonate).
C. Bleaching powder (calcium hypochlorite).
D. Vinegar (ethanoic acid).

*4

An atom of magnesium has 12 protons, 13 neutrons and 12 electrons. What is its mass (in atomic mass units)?
A. 12
B. 13
C. 24
D. 25

*5

Gloria Williams spilt some acid from a car battery (dilute sulphuric acid) on her kitchen table.

(a) Which of the following substances from her kitchen cupboard could she use to neutralise the acid?
A. Flour (mainly starch).
B. Salt (sodium chloride).
C. Vinegar (ethanoic acid).
D. Washing soda (sodium carbonate).

(b) Battery acid is fairly strong. Neutralising it in this way could produce a very lively reaction. It could get very hot, bubble and spit acid. What should Gloria do to prevent this?
A. Add less neutralising chemical.

B. Add more neutralising chemical.
C. Dilute the acid with cold water before using the neutralising chemical.
D. Dilute the acid with hot water before using the neutralising chemical.

*6

Astatine was discovered in 1940 in the USA. It has the following properties:
(i) It is solid at room temperature.
(ii) It does not conduct electricity or heat.
(iii) It does not react when it is heated in air.
(iv) The compound it forms with sodium has the formula NaAs.
The properties of astatine are similar to those of:

A. carbon.
B. iodine.
C. phosphorus.
D. sulphur.

*7

The cliffs at Kettleness near Whitby on the North Yorkshire coast contain alum shale. In the eighteenth century alum shale was used to make ammonia alum. Ammonia alum is a double salt of aluminium sulphate and ammonium sulphate

$$(NH_4)_2SO_4 \cdot Al_2(SO_4)_3 \cdot 24H_2O$$

which is used to fix dyes to fabrics. When heaps of alum shale are burned, a mixture of aluminium sulphate, $Al_2(SO_4)_3$, and iron (III) sulphate, $Fe_2(SO_4)_3$ is produced. The alum workers of Whitby had to convert the burnt shale to ammonia alum.

Which of the following could have solved their problem? (Hint: iron (II) carbonate does not dissolve in water.)

A. Adding stale urine (containing sodium chloride and ammonium carbonate).
B. Adding sea water (containing sodium chloride and magnesium chloride).
C. Adding powdered limestone (calcium carbonate).
D. Adding stale beer (containing ethanoic acid).

*8

The first stage in manufacturing sulphuric acid is to make the gas sulphur dioxide.

$$S(s) + O_2(g) \rightarrow SO_2(g)$$

A sulphuric acid works has about 25 brick furnaces where sulphur is burned in air. Each furnace uses 320 kg of sulphur every day. The process works best if 2/3 of the oxygen supplied is turned into sulphur dioxide.

(a) How much oxygen must be supplied to each furnace every day?
(Relative atomic masses O = 16; S = 32)
A. 210 kg.
B. 240 kg.
C. 320 kg.
D. 480 kg.

(b) The mass of air used must be
A. two-thirds as much as the answer to (a).
B. the same as the answer to (a).
C. four times as much as the answer to (a).
D. five times as much as the answer to (a).

*9 Carbon dioxide problem

Some scientists are worried that too much carbon dioxide is being produced. Carbon dioxide in the atmosphere stops heat escaping from the Earth and if the amount of carbon dioxide keeps increasing, the surface of the Earth could get very hot.

(a) The release of a large amount of carbon dioxide into the atmosphere could
A. reduce the ozone layer.
B. produce acid rain.
C. increase the greenhouse effect.
D. cause a nuclear ice age.

(b) Which of the following processes does not increase the level of carbon dioxide in the atmosphere?
A. Animal respiration.
B. Photosynthesis.
C. Combustion of fossil fuels.
D. Plant respiration.

(c) Some scientists believe that the release of too much carbon dioxide into the atmosphere could change the weather on Earth. There could be more storms and hurricanes, for example, or the Gulf Stream current which brings warm sea water from the equator to the British Isles could stop. This could happen because

    A. carbon dioxide traps energy in the atmosphere.

    B. carbon dioxide increases the amount of energy escaping from the atmosphere.

    C. carbon dioxide uses up energy in the atmosphere.

    D. solid carbon dioxide ('dry ice') falls as hail stones.

(d)

Fuel	Heat produced by burning 1 kg of fuel (MJ)	Volume of carbon dioxide produced by burning 1 kg of fuel (m³)
Coal	35	3.59
Natural gas	55	1.40

For the same amount of heat

    A. coal produces four times as much carbon dioxide as natural gas.

    B. natural gas produces four times as much carbon dioxide as coal.

    C. coal produces 1.6 times as much carbon dioxide as natural gas.

    D. natural gas produces 1.6 times as much carbon dioxide as coal.

*10

Mary Jones knows that bleaching is caused by chlorine. The stronger the bleach the more chlorine it releases.

(a) Bleach makes ink become colourless. Explain how Mary could use ink to find out the strength of a bleach. Assume that she has the usual laboratory equipment available. (3)

(b) Mary was asked to do a 'best buy' survey of bleaches on sale in the local supermarkets. Explain how she would use the results from the experiment above to find out which bleach gives best value for money. (3)

(c) Chlorine is a poisonous gas but it is added to drinking water. Explain this statement. (3)

11

A group of pupils have been given a sample of each of ten different types of coal. They have been asked to find out which is the best coal. They have decided to burn each coal and to find out how much ash (solid residue) is left.

(a) List the apparatus which they should use in doing the experiment. (4)

(b) Describe how they should do the experiment. (Include in your answer any safety precautions which they should take.) (5)

(c) How could they make sure that the experiment was complete? (3)

(d) Explain how the pupils could use their results to decide which is the best coal. (4)

12

(a) Vinegar is a solution of ethanoic acid in water. Food regulations require that vinegar sold in shops contains 50 g of ethanoic acid per dm³ (litre) of water. You are required to devise an experiment to check that the vinegar sold in your local supermarket meets the food regulations.

    Assume that the normal laboratory apparatus and common chemicals are available.

(b) The diagram shows the label from a bottle of Brown Sauce.

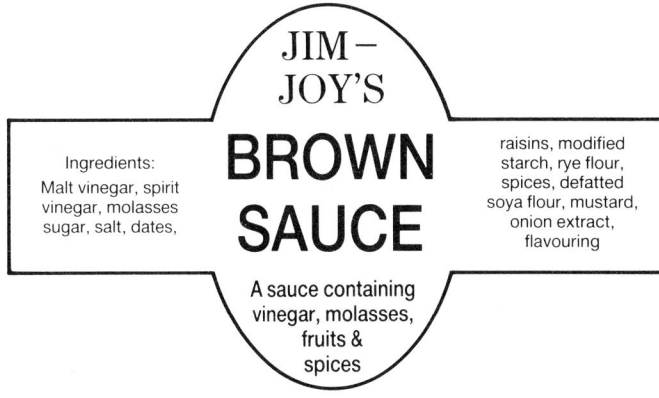

JIM – JOY'S

BROWN SAUCE

Ingredients: Malt vinegar, spirit vinegar, molasses sugar, salt, dates, raisins, modified starch, rye flour, spices, defatted soya flour, mustard, onion extract, flavouring

A sauce containing vinegar, molasses, fruits & spices

    (i) What modifications would you have to make to the experiment in (a) to find out how much vinegar there is in the Brown Sauce?

    (ii) Explain how you would find out the amount of solids in the Brown Sauce.

*13

The label below is on a jar of raspberry jam.

HARVEST JAM

340 g

INGREDIENTS

The following were used in making this jam:

Raspberries 156 g
Sugar 224 g
Citric acid
Gelling agent

Contains no artificial colours

(a) Which of the following techniques would you use to check that the jam contains no artificial colour?
 A. Chromatography.
 B. Distillation.
 C. Evaporation.
 D. Filtration. (1)

(b) The masses stated on the label do not seem to add up. What is the problem? (2)

(c) One of the jobs of a public analyst is to check that the jam does not contain too much water. Design an experiment that could be used to find out how much water the jam contains. (4)

*14

(a) Baking soda can be used to make baking rise. Its chemical name is sodium hydrogencarbonate. The equation for the effect of heat on sodium hydrogencarbonate is given below:

sodium hydrogencarbonate → sodium carbonate
 + water + carbon dioxide.

 (i) Which product from the action of heat on baking soda is responsible for a cake rising during baking? (1)

 (ii) Sodium carbonate has a bitter taste. Baking soda is not used for making plain sponge cakes but can be used to make chocolate or ginger cakes. Suggest why. (2)

(b) Another raising agent is baking powder, which consists of sodium hydrogencarbonate, tartaric acid and starch.

 (i) When water is added to baking powder, it immediately bubbles and froths. Explain what is happening. (2)

 (ii) Suggest why baking powder is used instead of baking soda to make a plain sponge cake. (1)

(c) (i) Explain why baking powder is sold in metal or plastic containers whereas baking soda is sold in cardboard containers. (3)

 (ii) Self raising flour contains baking powder. Explain why you should store it in an airtight container rather than in the paper bag in which it was sold. (2)

15

Calcium oxide is used to make calcium hydroxide which is the cheapest industrial alkali. At one time, calcium oxide was used to make mortar for building. Cement is now more popular for building.

Calcium oxide is made by roasting calcium carbonate (limestone).

$$CaCO_3 \rightarrow CaO + CO_2$$

This reaction works best at temperatures above 800 °C.

(a) The diagram shows a lime kiln at Jack Scout, Silverdale.

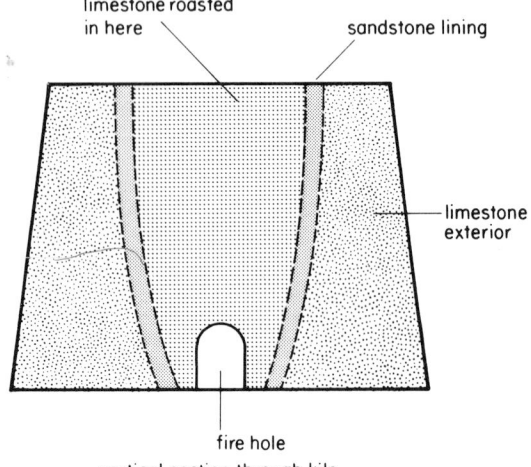

limestone roasted in here

sandstone lining

limestone exterior

fire hole
vertical section through kiln

Why is it lined with sandstone (silicon dioxide)? (2)

(b) Limestone is used to make iron in a blast furnace. As it is heated, the limestone first turns into calcium oxide. The calcium oxide then reacts with earthy impurities (silicon dioxide) in the iron ore to form slag.

$$CaO + SiO_2 \rightarrow CaSiO_3$$

The temperature in a blast furnace is above 1000 °C. At what temperature do you think the Silverdale lime kiln was operated? Explain your answer. (3)

*16 Loo cleaner

The diagram shows the label from a packet of Jim-Joy's toilet cleaner.

JIM – JOY'S
LOO
CLEANER

Kills germs

Removes lime scale

Jim-Joy's toilet cleaner is a powder in which the active ingredient is sodium dihydrogenphosphate (NaH_2PO_4).

(a) (i) When sodium dihydrogenphosphate is dissolved in water the solution has a pH of about 3. What does this tell you about the solution? *(1)*

(ii) Suggest how the solution removes lime scale from the toilet. *(3)*

(b) How could you prove that the toilet cleaner kills germs? *(4)*

The label on the back of the packet gives the following instructions:

A. Do not use with liquid bleach or other powder and liquid lavatory cleaners.

B. Use only for lavatory bowls.

(c) When an acid is added to bleach, chlorine is produced.

(i) Why is instruction **A** given? *(2)*

(ii) Suggest why Jim-Joy's loo cleaner should not be used with other lavatory cleaners. *(2)*

(iii) Give a possible reason why it should be used only for lavatory bowls. *(1)*

(d) Jim-Joy's toilet cleaner is poisonous. If a child swallows some of Jim-Joy's toilet cleaner something must be given to overcome its poisonous effect.

(i) Which of the following common household substances might a doctor recommend to be given to the child?

A. Baking soda (sodium hydrogencarbonate).

B. Epsom salts (magnesium sulphate).

C. Salt (sodium chloride).

D. Vinegar (ethanoic acid).

Give a reason for your answer. *(3)*

(ii) A doctor would usually give the child something to make the child be sick. Suggest why this is the best treatment for a child who has swallowed Jim-Joy's toilet cleaner. *(2)*

17 Making coffee

'Fresh' coffee is made using ground coffee beans and hot water. Jean uses a filter jug like the one in diagram 1. Sinita has bought an automatic filter machine like the one in diagram 2.

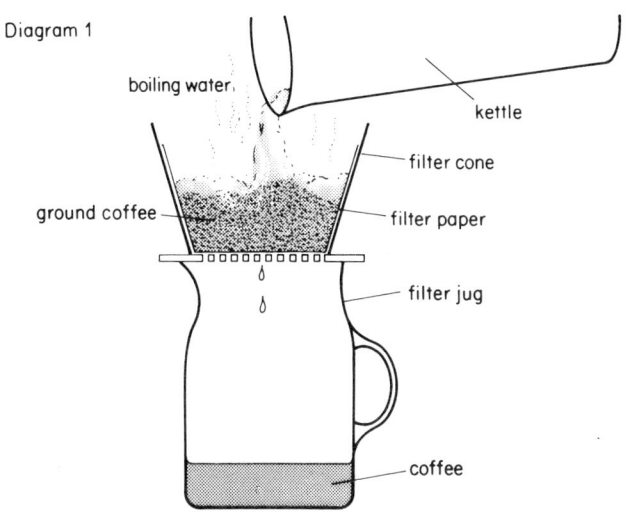

Diagram 1

boiling water

kettle

filter cone

ground coffee

filter paper

filter jug

coffee

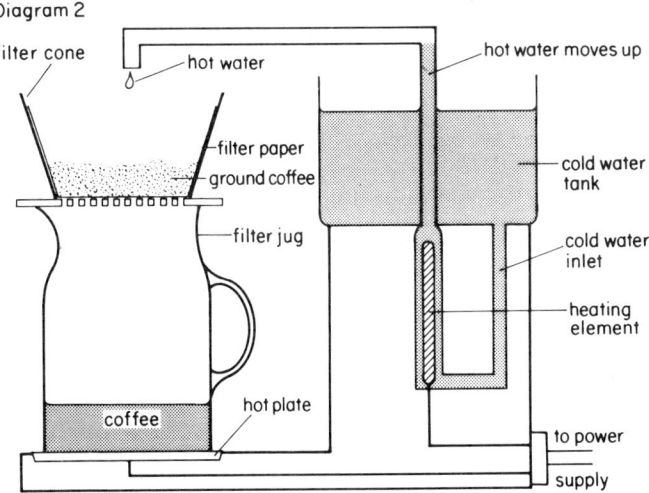

Diagram 2

filter cone

hot water

hot water moves up

filter paper

ground coffee

cold water tank

filter jug

cold water inlet

heating element

hot plate

coffee

to power supply

Sinita and Jean decided to compare their coffee makers. They used fine ground coffee for the first test.

(a) (i) Explain what they should keep the same to make sure that their test is fair. *(3)*

(ii) Suggest and explain two differences which they might notice between making coffee with the machine and with the filter jug. *(2)*

(b) In a second test the two girls used medium ground coffee. This has larger particles than fine ground coffee.

(i) Suggest and explain two differences that you would expect in the girls' results for this experiment compared to the first one. *(2)*

(ii) The girls want to get as many cups of 'dark' coffee as possible from 250 g of ground coffee beans. The diagram below shows a grinder in a supermarket. Which setting should they use? *(1)*

Diagram 3

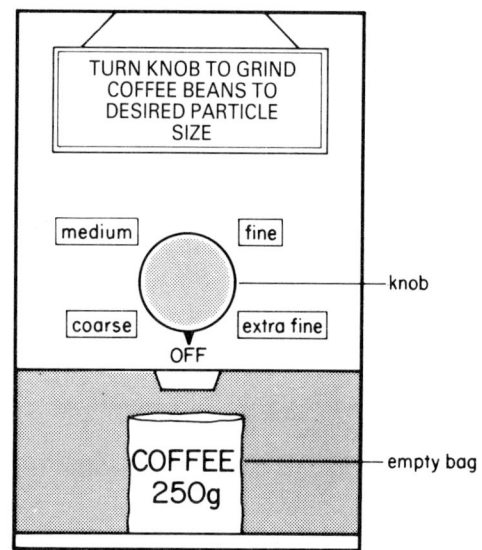

(c) The hotplate in Sinita's filter machine keeps the coffee hot. One day Sinita went out and left the machine switched on. When she came back, there was no liquid left in the jug, but its inside surface was coated with a dark brown powder.
(i) Explain carefully what has happened while Sinita was out. *(4)*
(ii) Describe one safety device that should be fitted to an automatic filter machine. *(2)*

18

Below is a map of a small town. The prevailing winds are from the west or south west.

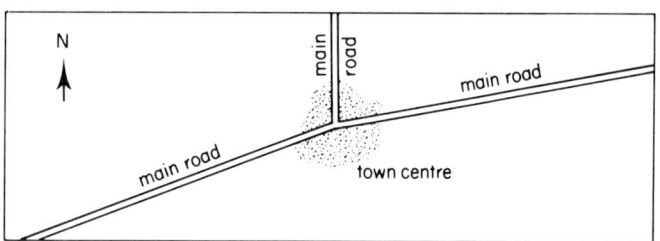

(a) (i) Which area is likely to have the highest amount of lead in the atmosphere? *(1)*
(ii) Give a reason for your answer. *(1)*
(b) Copy out the map and mark on it where you would site
(i) an industrial estate.
(ii) market gardens for growing salad vegetables. Give reasons for your choice of sites. *(4)*

*19 Solvay Tower

Ammoniacal brine is salt water in which ammonia has been dissolved (NH_3 + NaCl + H_2O).

Diagram 1 shows an outline of a Solvay Tower where ammoniacal brine reacts with carbon dioxide to produce sodium hydrogencarbonate.

Diagram 2 shows how the temperature varies in different parts of the Solvay Tower.

Diagram 1   Diagram 2

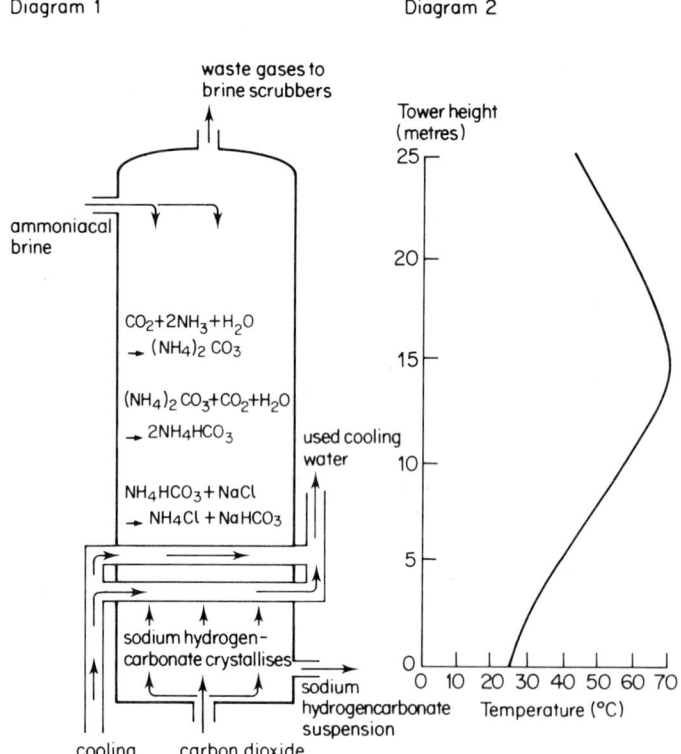

(a) (i) What is meant by the terms exothermic and endothermic when they are used to describe chemical reactions? *(2)*
(ii) The diagram shows the equations for three chemical reactions which take place in the Solvay Tower. Taken together, are they exothermic or endothermic? Give a reason for your answer. *(2)*
(b) The table gives information about the solubility of carbon dioxide and sodium hydrogencarbonate in hot and cold water.

Substance	Solubility (grams of substance in 1 litre of water)	
	Cold water	Hot water
Sodium hydrogencarbonate	95 (at 20 °C)	142 (at 100 °C)
Carbon dioxide	3.3 (at 20 °C)	0.7 (at 60 °C)

Explain why the liquid at the bottom of the Solvay Tower must be cooled for the processes to work properly. *(4)*

(c) Which substances are in the waste gases that leave the top of the tower? *(2)*

(d) The temperature in the Solvay Tower is controlled carefully to give the correct size of sodium hydrogencarbonate crystals for filtering out of the suspension.
   (i) What is a 'suspension'? *(1)*
   (ii) Explain how the size of sodium hydrogencarbonate crystals could be related to the filtration process. *(2)*
   (iii) Describe how you could produce
      1. small crystals,
      2. large crystals,
      from a solution of sodium hydrogencarbonate. *(4)*

20   Smog

Normally the air at ground level in a town is warmer than the air a few hundred metres above the town. Hot air rises because it is less dense than cold air. So gases from car exhausts and the burning of coal or oil in factories and homes tend to rise away from the ground. But sometimes there is a 'temperature inversion' in the atmosphere. When this happens, air a hundred metres or more above ground is at a higher temperature than the air at ground level.

The graphs below show how air temperature can change with height above ground level. All diagrams are for a day when there are no clouds in the sky.

Diagram 1

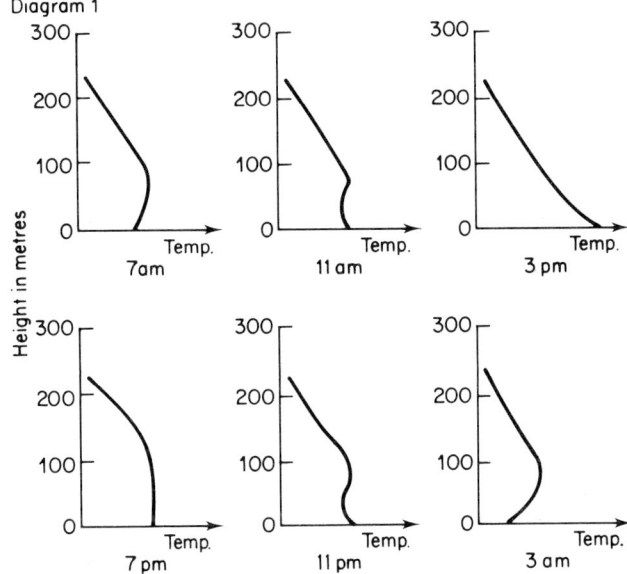

Many small towns in East Lancashire are in valleys. Diagram 2 shows how smog built up on a sunny April day in the 1950s.

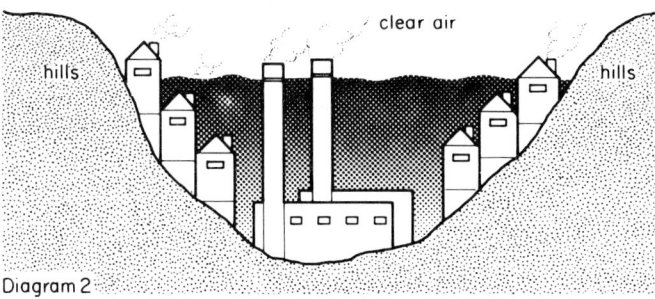

Diagram 2

(a) (i) What is the height of the temperature inversion shown by the graphs? *(1)*
   (ii) At what time of day does the temperature inversion start? *(1)*
   (iii) Estimate the time at which the temperature inversion disappears. *(1)*
   (iv) What time of day do you think it was in the town shown in diagram 2? *(1)*

(b) Suggest why this kind of temperature inversion happens on cloudless days. *(2)*

(c) How do tall factory chimneys help solve the smog problem? *(1)*

(d) A common home improvement during the 1950s was a type of fire grate which made it easy to keep coal fires burning all night. How did this affect the smog problem? *(2)*

(e) (i) Many of the towns in East Lancashire became 'smokeless zones' during the 1960s. Why are 'smokeless zones' more important in valley towns than in towns built in flat areas? *(2)*
   (ii) There is not much smoke now. But temperature inversions can still damage people's health in valley towns. Why is this? *(2)*

*21   Tooth problems

Ben Holland went to the dentist for a check-up. The dentist said that he needed a filling in one tooth. She said that she would use a new alloy called Fillsome for the filling. The diagram shows the composition of Fillsome.

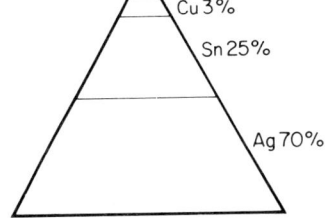

(a) (i) What is meant by the term alloy? *(1)*
    (ii) Which element is present in Fillsome in the greatest amount? *(1)*
(b) Suggest and explain three properties that Fillsome should have. *(6)*
(c) To prepare the Fillsome, the dentist mixes it with mercury.
    (i) Give the name of the type of compound which is formed when mercury is mixed with Fillsome. *(1)*
    (ii) Suggest one advantage and one disadvantage of using this mixture. *(2)*

The dentist told Ben that he would need fewer fillings if he cleaned his teeth with 'Fluoride toothpaste'. On his way home, Ben bought a 50 gram tube of Macduff's toothpaste. Its label said it contained

    0.8% w/w sodium monofluorophosphate
    0.1% w/w calcium glycerophosphate.

(d) Calculate the mass of each compound in Ben's tube of Macduff's toothpaste. *(4)*
(e) Ben found that he was able to clean his teeth 100 times before he needed to buy a new tube of Macduff's.
    (i) The formula for sodium monofluorophosphate is $NaFPO_4$. Calculate the mass of one mole of sodium monofluorophosphate. (Relative atomic masses: $Na = 23; F = 19; P = 31; O = 16$) *(3)*
    (ii) Calculate the 'dose' of sodium monofluorophosphate Ben's teeth receive at each cleaning. *(2)*
    (iii) How many moles of fluoride ions are contained in this 'dose'? *(2)*
    (iv) State one assumption which you have made in (ii). *(1)*

22 Water vapour

(a) George left two fish bowls on a window sill for a week. One contained fresh water. The other contained salty sea water.
This is how they appeared at the start of a week.

Diagram 1

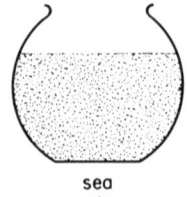 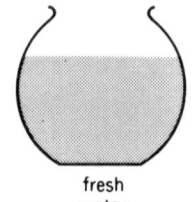

sea water   fresh water

This is how they appeared at the end of the week.

Diagram 2

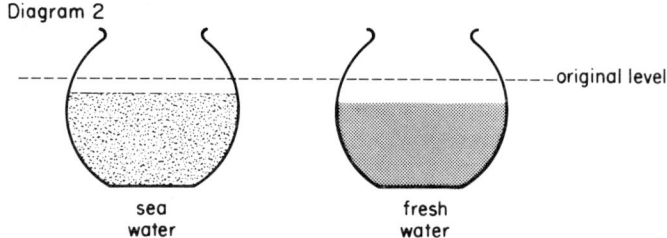

sea water   fresh water

Suggest why the levels change like this. *(2)*

(b) George decided to see what happens if the two bowls are kept in a sealed container.
This is how they appeared at the start of a week.

Diagram 3

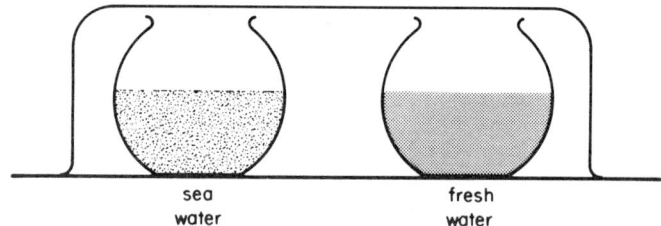

sea water   fresh water

This is how they appeared at the end of the week.

Diagram 4

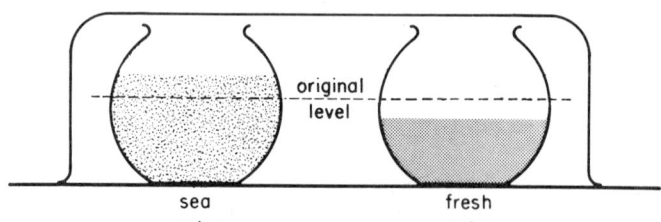

sea water   fresh water

Suggest why the levels change like this. *(3)*

(c)

Julia has damp patches on her bedroom wall even in dry weather. They tend to dry out whilst she is away on holiday. They are made worse by having plants on the window sill. Suggest an explanation for this problem and two ways of overcoming it. *(4)*

(d) Animal skins are used for making leather. The cells in the skin contain moisture. To stop the skins from rotting, this moisture must be dried out as soon as the skins are removed from the animals. In hot countries the skins are stretched out in the sun. In England salt is scattered on them. Suggest how the salt works. *(2)*

*23 Liming gardens

Ammonium sulphate is often added to soil as a nitrogen fertiliser. It can decompose:

$$(NH_4)_2SO_4(s) \rightarrow 2NH_3(g) + H_2SO_4(aq)$$

(a) Adding ammonium sulphate to soil changes the pH of the soil. Explain why. *(1)*
(b) Julia measured the pH of the soil in her garden. She read her gardening book which said that burnt lime (CaO) can be used to correct the pH to 7. Fifty g/m^2 is needed. However, the only two types of lime on sale at the Garden Centre were hydrated lime $(Ca(OH)_2)$ and ground limestone $(CaCO_3)$. The following equations show how each of the three types of lime react in the soil.

$$CaO + H_2SO_4 \rightarrow CaSO_4 + H_2O$$

$$Ca(OH)_2 + H_2SO_4 \rightarrow CaSO_4 + 2H_2O$$

$$CaCO_3 + H_2SO_4 \rightarrow CaSO_4 + CO_2 + H_2O$$

(Relative atomic masses: H = 1; C = 12; O = 16; Ca = 40)
(i) How much hydrated lime should be used per square metre? *(3)*
(ii) How much ground limestone should be used per square metre? *(2)*

24 Where to build a chemical works

The best site for a chemical works may depend on transport costs.
    In diagram 1, M represents the market and S the source of raw material for a simple process with only one raw material. Suppose that it costs £1 to transport a load of raw material 1 km and that it costs £1 to transport the product made from it 2 km.

Diagram 1

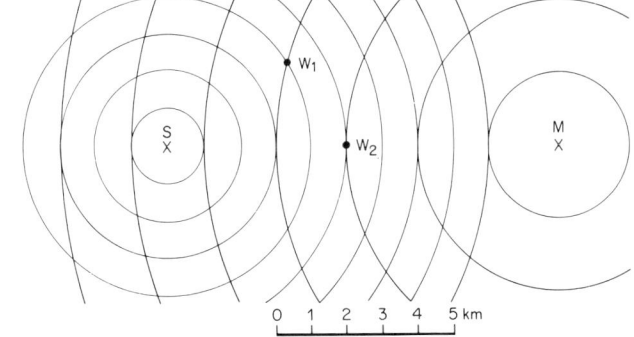

(a) (i) Why is it usually more costly to transport raw material than the product which is made from it? *(1)*
    (ii) What is the transport cost per load of raw material if the works is at W_1? *(1)*
    (iii) What is the cost per load of raw material if the works is at W_2? *(1)*
    (iv) Copy the diagram. On your diagram draw a line joining all the places where the cost of transport per load of raw material is £9. This line is called an isodapane. *(2)*
(b) Nearly a million people in Nairobi and the surrounding area use charcoal for cooking food and boiling water. Wood is the only raw material used in making charcoal. Charcoal made at Githumu has a transport cost of £5 per tonne by the time it gets to Nairobi.

Diagram 2

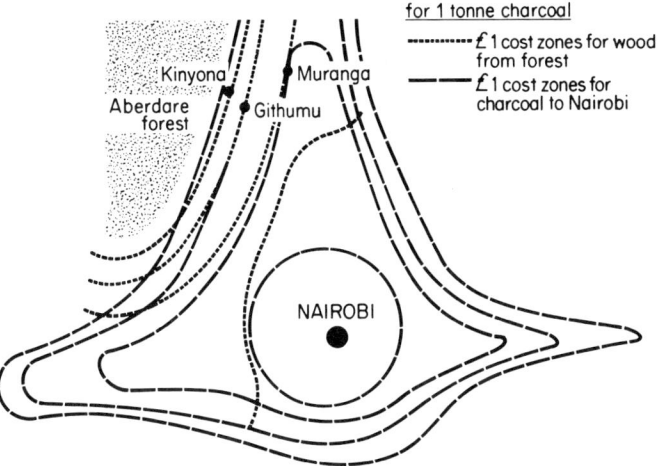

(i) What would be the transport cost on arrival at Nairobi of charcoal made at
    1. Kinyona? *(1)*
    2. Muranga? *(1)*
(ii) Why are the transport cost zones for carrying charcoal to Nairobi not always the same width? *(2)*

(iii) Draw a diagram showing the Aberdare forest and Nairobi. Show the main roads leaving Nairobi. *(2)*

(c) Technology has changed the iron and steel industry. Between 1900 and 1980 the amount of coal needed to make a tonne of steel fell by 70%. There has been a great increase in the use of scrap. This is readily available in the places where the products are used (the markets).

Diagram 3

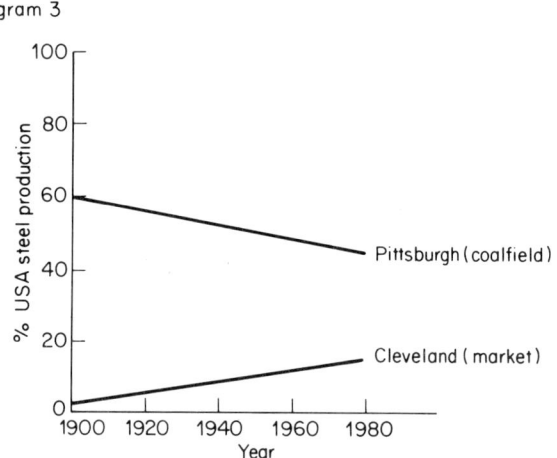

Explain why better technology has caused Pittsburgh a problem. *(3)*

(d) Low grade iron ore extracted in Minnesota, USA, is converted into pig iron in nearby Duluth. It is then transported 800 km to Chicago. High grade iron ore from the north of Sweden is transported to the Stockholm area 800 km away. It is then converted into pig iron for use nearby.
  (i) Explain these two solutions to the transport cost problem. *(4)*
  (ii) Suggest two other factors which might affect the decision about where to manufacture pig iron. *(2)*

---

25 Enriching uranium

At Capenhurst near Chester, uranium is enriched for re-use in nuclear power stations. The fissile isotope used in nuclear reactors is U-235. The heavier isotope U-238 is more common. The two isotopes are separated in a set of centrifuges which work like spin-dryers. Heavy material is flung to the outside of the drum. Lighter material falls back to the centre of the drum.

(a) Solid particles of uranium compounds contain both molecules made from U-235 and molecules made from U-238. But the centrifuges cannot work on solids.

Liquids and gases have molecules which can move freely. The compound of uranium used for this separation is uranium hexafluoride (UF_6). Tables of boiling and melting points give

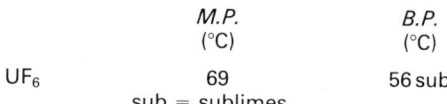

	M.P. (°C)	B.P. (°C)
UF_6	69	56 sub

sub = sublimes

How can suitable operating conditions be provided? *(1)*

(b) The diagram shows where U-235 and U-238 uranium hexafluoride are found in the centrifuge.

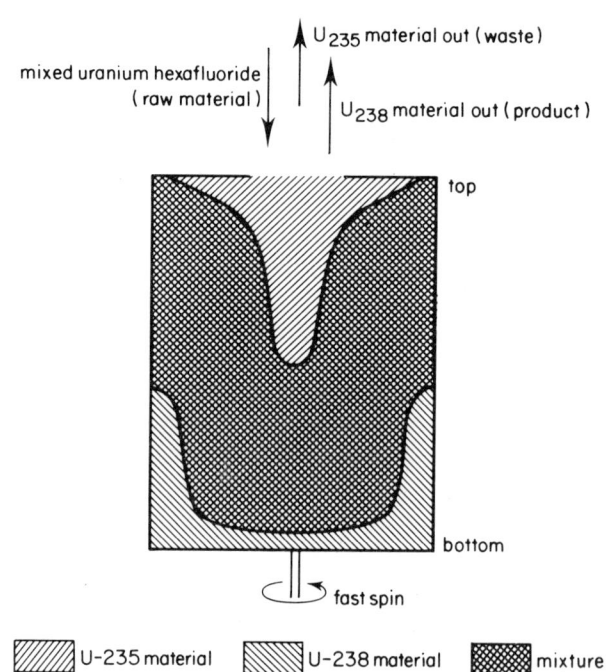

  (i) Explain why U-238 hexafluoride is found mostly at the bottom. *(2)*
  (ii) Draw another diagram of the centrifuge. It should have a closed top and suitable pipes to get the raw material in and products and waste out. *(4)*

---

26 Insecticide problem

(a) Garlic oil can kill housefly larvae. Some 'organic' gardeners squeeze tomatoes in water and use this water to spray cabbages. This reduces damage by the caterpillars of 'cabbage white' butterflies.

How do you think people first found out that garlic and tomatoes could be used in this way? *(3)*

(b) Until 1940, nicotine from tobacco plants and pyrethrin from pyrethrum flowers were widely used as insecticides. Pyrethrum flowers were grown for this purpose as a crop on large farms in Kenya. But crop failure and the cost of transport were problems. Synthetic pyrethrin is now made in laboratories. Pyrethrin is not very poisonous to larger animals.

The chemical DDT was first made in 1874. In 1939, at the the start of World War II, it was found that DDT is a very effective insecticide. It is not poisonous to larger animals and it can be made cheaply.

(i) Crops can fail because of
drought,
lack of sun,
fungus attack (e.g. 'rust' or 'black spot'),
insect attack (e.g. plagues of locusts),
or because workers are away at war and unable to harvest the crop.
Which one of these is unlikely to be a problem with pyrethrum? (1)

(ii) Very little pyrethrum was being grown in Kenya by the 1960s. Suggest reasons for this. (4)

(c) DDT has been used to kill mosquitoes. But some mosquitoes have an enzyme called 'DDTase' which can deactivate it. As these mosquitoes breed, new resistant strains of mosquito develop.

Diagram 1

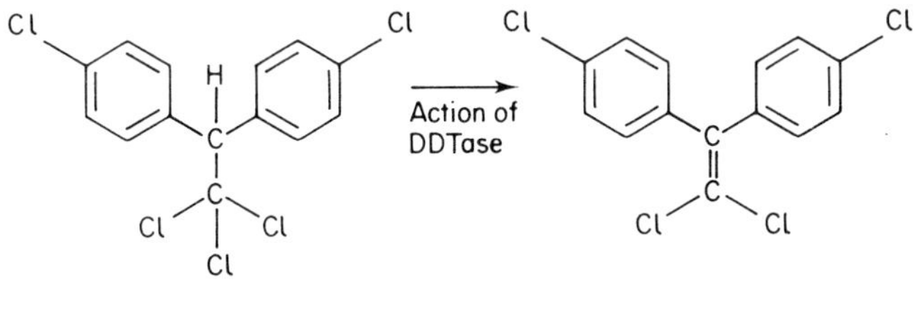

DDT   DDE

What compound is removed from DDT as DDTase converts it to DDE? (1)

(d) DDT is not soluble in water but it is soluble in fat. Therefore:
1. it builds up in the body fat.
2. it cannot be excreted from the body in urine.
3. it is not biodegradable, so land treated with DDT is 'poisoned' for many years.

Methoxychlor is an insecticide which is soluble in water.

Diagram 2

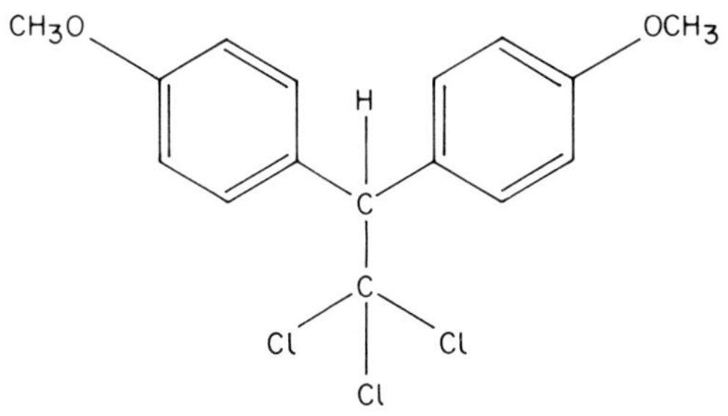

methoxychlor

(i) How does a methoxychlor molecule differ from a DDT molecule? (2)
(ii) Suggest and explain three properties that methoxychlor could have (see 1, 2, 3 above). (3)
(iii) Explain whether you would expect DDTase to deactivate methoxychlor. (2)

Diagram 3

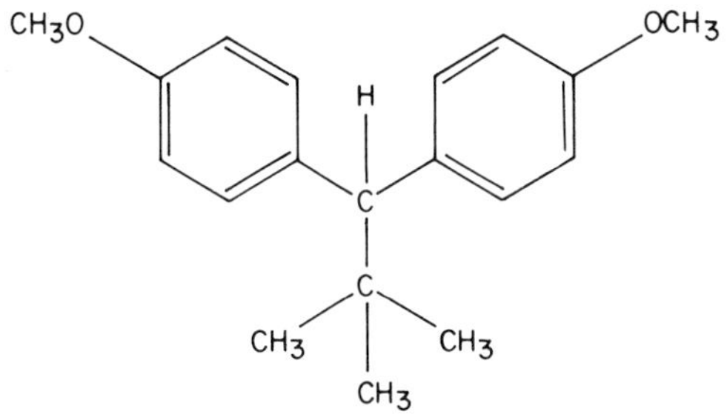

compound M

(i) In what way is the compound M molecule different from the methoxychlor molecule? (2)
(ii) Explain whether you would expect DDTase to deactivate it. (2)
(iii) Explain whether compound M would be biodegradable. (2)

(f) Piperonyl butoxide is not an insecticide. By itself it has no observable effect on insects or larger animals. When it is used together with some insecticides it increases their power. The increased effect when two chemicals are used together is called synergism. Suggest how piperonyl butoxide works. (2)

3 Patterns

1

These are the main steps in the carbon cycle:

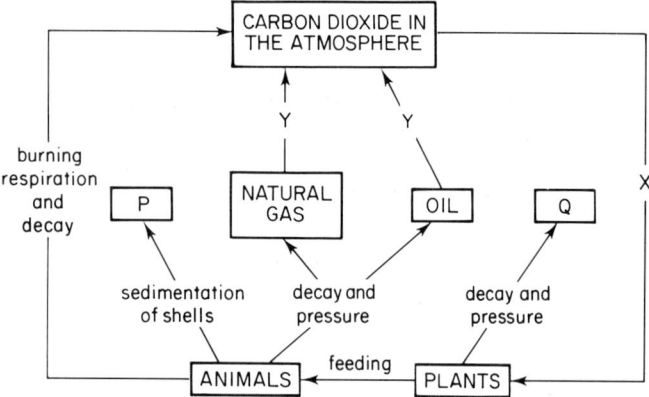

Use this diagram of the carbon cycle to answer questions (a) to (d). Choose your answers from this list:
A. Chalk and limestone
B. Coal
C. Burning
D. Photosynthesis
(a) What is happening at X?
(b) What is happening at Y?
(c) What is substance P?
(d) What is substance Q?

2

The accompanying diagram shows the main steps in the nitrogen cycle, with some gaps.

Study the diagram carefully and decide where these missing labels should go. (One has been done to help you.)

Missing label	Place it belongs
Taken up by roots	D
Nitric acid	
Haber process	
Bacteria at work	
Thunderstorms	
Legumes, e.g. peas and beans	

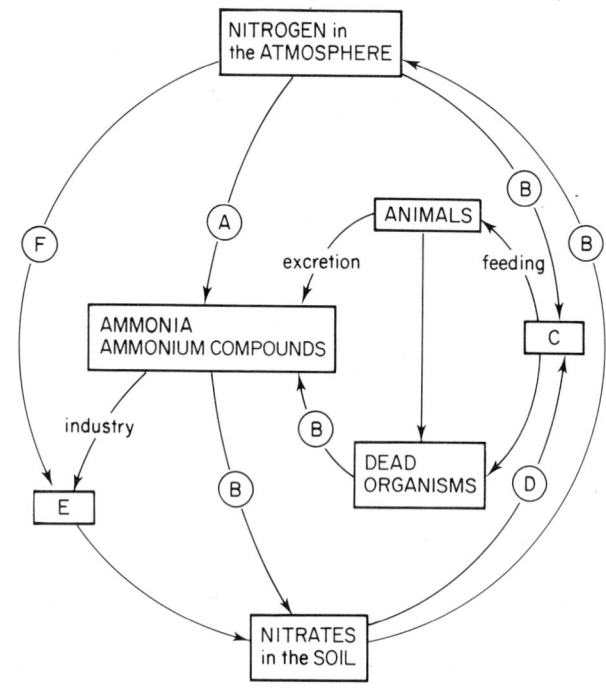

*3

The table summarises some information about four elements P, Q, R and S.

Element	Type	Charge on ions formed
P	metal	+2, +3
Q	non-metal	−2
R	metal	+2
S	non-metal	+3, +5, −3

(a) The conclusion which is supported by the evidence in the table is that
   A. metals form only positive ions.
   B. non-metals form only negative ions.
   C. metals form only one ion.
   D. non-metals form many ions.

(b) Which one of the following compounds is NOT possible?
   A. PQ
   B. RQ
   C. PS
   D. PR

(c) Which one of the elements in the table forms an atom with full inner shells and just two electrons in its outer shell?
- **A.** P
- **B.** Q
- **C.** R
- **D.** S

4

This table shows the distance between atoms in some bonds (bond length) and the energy needed to break the bond.

Bond	Bond length $(10^{-10}m)$	Energy needed to break bond (kJ/mol)	Atomic number
C-C	1.54	347	6
C=C	1.34	619	6
Cl-Cl	1.99	242	17
Br-Br	2.28	193	35
I-I	2.66	151	53

Based on the information in the table, which of the phrases below will finish this sentence correctly?
'The energy needed to break a bond is . . . .'
- **A.** lower the longer the bond.
- **B.** lower the shorter the bond.
- **C.** not affected by bond length.
- **D.** linked to atomic number.

5

Use the information in this table to answer questions (a) and (b).

Substance	Melting point (°C)	Boiling point (°C)	Electrical conductivity when molten
Ammonia	−77	−34	poor
Carbon dioxide	−111	−78	poor
Copper (II) chloride	620	990	good
Lead (II) chloride	501	950	good
Methylated spirits	−100	80	poor
Sodium chloride	801	1413	good
Sugar	184	decomposes	poor
Silicon dioxide	1610	2230	poor

(a) Which of the following patterns is shown in the table?
- **A.** Good conductors have high melting points.
- **B.** Good conductors have low melting points.
- **C.** Poor conductors have low melting points.
- **D.** Poor conductors have high melting points.

(b) Which of the following substances are liquid at 25 °C?
- **A.** Ammonia.
- **B.** Methylated spirits.
- **C.** Sodium chloride.
- **D.** Sugar.

6

In developing countries, agricultural officers ask farmers what crops already grow on their land. The officers can then advise the farmers which new crops to try growing. The agricultural officers use the information in diagrams 1 and 2.

Diagram 1
pH preferred by plants already growing

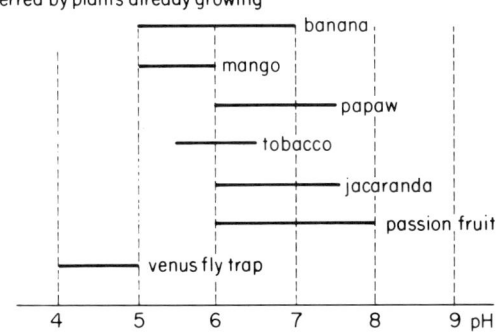

Diagram 2
pH preferred by possible new crops

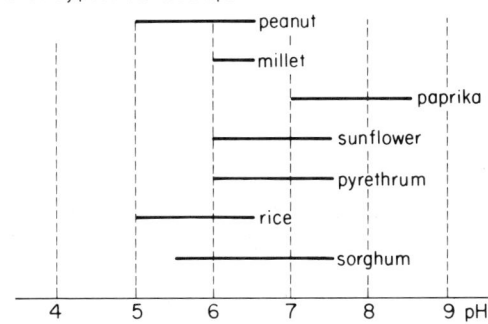

Complete each of the following pieces of advice that the agricultural officers are correct to give.
(a) Paprika could be grown if these plants grow well:
- **A.** Jacaranda, papaw, passion fruit.
- **B.** Banana, mango, tobacco.
- **C.** Banana, papaw, jacaranda.
- **D.** Banana, passion fruit.

(b) Millet will grow well if the land already has
   A. mango and tobacco.
   B. papaw and tobacco.
   C. banana and mango.
   D. Venus fly trap.
(c) If the land already grows mango well, it could grow
   A. millet.
   B. peanuts.
   C. pyrethrum.
   D. sunflowers.

7

'Itai-Itai Byo' is a disorder where calcium ions in the bones are replaced by ions of another metal. It was first found in Japan in the 1950s.

A calcium atom has two electrons in its outer shell. All other shells are full. The radius of the calcium ion is 100×10^{-12} m.

Calcium ions can be replaced most easily by a metal ion with similar chemical properties and similar size.

Metal	Number of electrons in outer shell	Other shells full?	Ionic radius (m)
A Sodium	1	yes	102×10^{-12}
B Cadmium	2	yes	95×10^{-12}
C Silver	1	yes	115×10^{-12}
D Zinc	2	yes	75×10^{-12}

(a) Which of these metals could replace calcium most easily?

(b) Strontium chloride is put into special toothpaste. The toothpaste is used by people whose teeth hurt when they eat things like oranges or ice cream. It makes teeth less sensitive. Strontium has two electrons in its outer shell. All other shells are full. Its ion has a radius of 113×10^{-12} m.

Which of the following is the most likely explanation of the way strontium chloride works?
   A. Strontium chloride neutralises acid in fruit juice.
   B. Strontium replaces calcium in teeth.
   C. Strontium replaces sodium in the nerves which detect pain.
   D. Strontium chloride forms a thick layer on teeth which prevents heat conduction.

*8

The diagram shows the fire risk for different types of cotton night clothes.

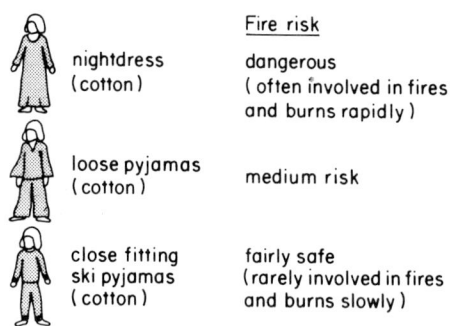

	Fire risk
nightdress (cotton)	dangerous (often involved in fires and burns rapidly)
loose pyjamas (cotton)	medium risk
close fitting ski pyjamas (cotton)	fairly safe (rarely involved in fires and burns slowly)

Give two possible explanations for the pattern of risk shown in the diagram.

*9

The table below shows the pattern of solubility of compounds formed from pairs of ions: soluble in water ($\sqrt{}$), and insoluble in water ($\times$).

(a) State which of the following compounds are soluble in water:

(i)   ammonium carbonate,
(ii) copper chloride,
(iii) iron hydroxide,
(iv) lead sulphate,
(v)   potassium nitrate.   (3)
(b) Write down two patterns shown in the table. (2)
(c) From the information in the table suggest, briefly, how you could make a small sample of magnesium carbonate.   (3)

				Cation				
Anion	ammonium	sodium	potassium	magnesium	zinc	copper	iron	lead
nitrate	$\sqrt{}$	$\sqrt{}$	$\sqrt{}$	$\sqrt{}$	$\sqrt{}$	$\sqrt{}$	$\sqrt{}$	$\sqrt{}$
chloride	$\sqrt{}$	$\sqrt{}$	$\sqrt{}$	$\sqrt{}$	$\sqrt{}$	$\sqrt{}$	$\sqrt{}$	$\times$
sulphate	$\sqrt{}$	$\sqrt{}$	$\sqrt{}$	$\sqrt{}$	$\sqrt{}$	$\sqrt{}$	$\sqrt{}$	$\times$
carbonate	$\sqrt{}$	$\sqrt{}$	$\sqrt{}$	$\times$	$\times$	$\times$	$\times$	$\times$
hydroxide	$\sqrt{}$	$\sqrt{}$	$\sqrt{}$	$\times$	$\times$	$\times$	$\times$	$\times$

10

The table below shows how the energy content of different types of fuel varies with carbon content.

Fuel	Mass of carbon in 1 kg of fuel (kg)	Energy content of 1 kg of fuel (kJ)
Wood	0.50	20 000
Peat	0.60	19 000
Lignite	0.62	25 000
Brown coal	0.70	27 200
Bituminous coal	0.79	32 000
Anthracite	0.91	33 000

(a) What pattern links the carbon content of the fuel to its energy content? (2)
(b) If the fuel is burned with enough air for complete combustion, which fuel would leave the least ash? Explain your answer. (4)

11 Ammonia

Ammonia is made industrially by reacting nitrogen and hydrogen.

$$N_2(g) + 3H_2(g) \rightarrow 2NH_3(g)$$

The reaction does not go to completion: there is always some nitrogen and hydrogen left.

The table below shows the volume (m^3) of ammonia gas produced when 2 m^3 of nitrogen is reacted with 6 m^3 of hydrogen.

Temperature (°C)	Pressure (atmospheres)				
	25	50	100	200	400
100	3.82	3.90	3.92	3.96	3.98
200	3.12	3.38	3.60	3.76	3.88
300	1.92	2.26	2.76	3.20	3.54
400	0.64	1.06	1.62	2.24	2.86
500	0.22	0.42	0.76	1.24	1.94

(a) (i) What happens to the volume of ammonia produced at 500 °C as the pressure is increased? (1)
(ii) Is the pattern the same for other temperatures?
Answer yes or no. Explain your answer. (2)
(b) (i) What effect does increasing the temperature have on the volume of ammonia produced at a pressure of 100 atmospheres? (1)
(ii) Is the pattern the same for the other pressures? Answer yes or no. Explain your answer. (2)
(c) Which temperature and pressure produce the greatest volume of ammonia? (2)
(d) For 200 atmospheres pressure, plot a graph of temperature (vertical axis) against the volume of ammonia produced (horizontal axis). (4)
(e) Industry uses a temperature of 450 °C and a pressure of 200 atmospheres.
(i) How much ammonia will be produced from 2 m^3 of nitrogen and 6 m^3 of hydrogen at this temperature and pressure? (1)
(ii) How much ammonia would be produced if a temperature of 100 °C had been used at the same pressure? (1)
(iii) Give a reason why industry uses a higher temperature. (1)
(f) The industrial process also uses a catalyst. The catalyst does not alter the amount of ammonia produced. Explain why a catalyst is used. (1)

*12 Metals

In an experiment pupils added small samples of metal powder to five different solutions. They repeated the experiment until they had tested five different metals with fresh samples of the solutions. The table below shows the results of their experiment.

Solution containing ions of	Metal sample added				
	Calcium	Magnesium	Zinc	Iron	Copper
Calcium	×	×	×	×	×
Copper	√	√	√	√	×
Iron	√	√	√	×	×
Magnesium	√	×	×	×	×
Zinc	√	√	×	×	×

√ = the metal in solution is displaced
× = nothing happens

(a) Use the results in the table to put the five metals in order, most reactive first. (2)
(b) For homework, the pupils were given the results for the same experiment when metals X and Y were added to the solutions:

	X	Y
Calcium	×	√
Copper	×	√
Iron	×	√
Magnesium	×	√
Zinc	×	√

(i) Put X and Y in the correct places in the list you gave in (a). *(2)*

(ii) What could X and Y be? *(2)*

(iii) For each of X and Y, suggest why the pupils were not asked to use the metal in their class experiment. *(4)*

(c) (i) Write an equation to describe the reaction which happens when zinc is added to the solution containing iron (II) ions. *(3)*

(ii) Describe one way in which this reaction can be used to make water tanks last longer. *(3)*

(d) Some ships have a large ball of magnesium which hangs over the side in the water. Explain what you think its purpose is. *(3)*

(e) Nuggets of gold have sometimes been found in Wales. Some of Queen Victoria's jewellery was made from Welsh gold. There were copper mines operating in Wales three hundred years ago. When the mines were first opened, the water running in the mines was clear. After miners had been working with picks and shovels for a few weeks, the water running in the mines was found to contain glittering orange coloured particles. The miners thought it was gold. But Robert Boyle, a well known scientist at that time, explained that it was something else. What explanation do you think he gave? *(3)*

A ⋆13 Fuels

The table below shows the energy which is given out when four hydrocarbon fuels are burned in air.

Fuel	Energy given out (kJ/mol)
CH_4	890
C_2H_6	1560
C_3H_8	2220
C_4H_{10}	2877

(a) (i) Describe the pattern linking size of molecule to the energy it gives out. *(2)*

(ii) Suggest a reason for this pattern. *(2)*

(b) (i) Explain why these compounds are all called hydrocarbons. *(2)*

(ii) Give the name of the family of hydrocarbons to which they all belong. *(1)*

(iii) Draw the structural formula for the first two compounds in the table. Under each one write its name. *(4)*

(c) (i) Estimate the energy given out by hydrogen when it burns. *(2)*

(ii) Give a reason, other than energy output, why CH_4 is preferred to hydrogen as the domestic fuel for cooking. *(1)*

14 Petrol

The 'octane number' of a petrol tells you how smoothly an engine will run when using it. Petrol with a low octane number explodes very fast. It burns very inefficiently. The engine 'pinks' or 'knocks'.

Alkanes can be straight chained

or branched:

Alkenes have a double bond:

The accompanying table shows the structure of a number of hydrocarbons and their octane number.

Hydrocarbon	Molecular formula	Structure	Octane number
Octane	C_8H_{18}		−19.0
Heptane	C_7H_{16}		0.0
Hexane	C_6H_{14}		24.8
Pentane	C_5H_{12}		61.7
Butane	C_4H_{10}		93.8
Propane	C_3H_8		97.1
Oct-2-ene	C_8H_{16}		56.2
Oct-4-ene	C_8H_{16}		73.3
Hex-2-ene	C_6H_{12}		92.7
Pent-1-ene	C_5H_{10}		90.9
But-1-ene	C_4H_8		97.4
2,4-dimethylhexane	C_8H_{18}		65.2
2,2,4-trimethylpentane	C_8H_{18}		100.0

(a) What patterns can you see for
   (i) straight chain alkane molecules? (2)
   (ii) straight alkene molecules? (2)
(b) What effect does branching have on octane number? (3)

15

At one time, relative atomic masses (RAM) were found by comparison with hydrogen or oxygen. RAM was then known as Atomic Weight. The table of International Atomic Weights for 1921 lists 81 elements. Data for all those starting with letter A are given below.

1921 Atomic Weights (Relative Atomic Masses)

Element	Atomic Weight	
	H = 1 (compared with hydrogen)	O = 16 (compared with oxygen)
Aluminium	26.8	27.1
Antimony	119.2	120.2
Argon	39.6	39.9
Arsenic	74.37	74.96

Today, the element carbon is used as the standard element against which the masses of other atoms are compared. Carbon atoms are given a Relative Atomic Mass of 12. The 1983 International Relative Atomic Mass table lists 107 elements. Data for all those starting with the letter A are given below.

1983 Relative Atomic Masses, reliable to ±1 in last digit (* Wide range in quantities of different isotopes in natural samples makes it impossible to give a more accurate value.)

Element	RAM
Actinium	227.0278
Aluminium	26.98154
Americium	243.0614
Antimony	121.75±3
Argon	39.948*
Arsenic	74.9216
Astatine	209.9871

(a) Why do you think actinium, americium and astatine are missing from the 1921 list? (2)

(b) The 1921 table gives the Atomic Weight of oxygen as 15.87 in the left hand column (compared to hydrogen). What do you think the number is in the right hand column? (1)

(c) There is a pattern linking the 1921 figures which compare elements to hydrogen and those which compare elements to oxygen. But it is not a perfect pattern.
   (i) Describe the pattern. (2)
   (ii) Why do you think the pattern is not perfect? (2)
   (iii) The 1921 figure for uranium is 236.3 compared to hydrogen. Use the pattern to estimate what it should be compared to oxygen. (1)

(d) What do you think the 1983 table gives as the Relative Atomic Mass of hydrogen (compared to carbon)? (2)

Classifying elements

(Some of the background to this question is discussed on page 59, the introduction to answers on Patterns.) This is a long question. You may not want to do it all at one time.

(a) Metals

It is sometimes said that:
metals are *malleable* (they can easily be shaped by rolling or beating), non-metals are not.
metals are *lustrous* (shiny), non-metals are not.
metals are *dense*, non-metals are not.
metals are *electrical conductors*, non-metals are not.

   Here is some information about a number of elements. Are there exceptions to the patterns stated above?

Aluminium is a white metal with a blue tinge and density varying from 2703 kg/m^3 (cast) to 2709 kg/m^3 (rolled). Its electrical conductivity is 40 units and it is used for electric cables in some countries.
Bismuth is a white metal with a reddish tinge. Its density is 9803 kg/m^3. It is brittle and easily powdered. It has a melting point of 271 °C and is used in making very low melting point alloys: Wood's metal (71 °C) and Lipowitz' alloy (60 °C). Its electrical conductivity is 0.93 units.
Amorphous *Carbon* has a density of 2261 kg/m^3 and its electrical conductivity is 0.33 units. Some forms of carbon, coke and diamond, are shiny. It can be formed into pencil leads and threads for carbon fibre reinforced boats. The first electric lamps had carbon filaments. Diamond is the hardest material on Moh's hardness scale.
Gold is a bright and shiny yellow metal. It can be beaten into leaves only 0.0005 mm thick. The thinnest layers transmit green light. Its density is 19 281 kg/m^3 and its electrical conductivity is 48 units.
Iodine is a blackish-grey crystalline solid with a metallic lustre. Its density is 4953 kg/m^3. It melts at 114 °C and boils at 184 °C to form a purple vapour. It does not conduct electricity (electrical conductivity 10^{-19} units).
Sodium is a silver-white soft metal. It cuts easily with a knife and the cut surface tarnishes rapidly in air. It has a dim greenish glow in the dark. Its electrical conductivity is 23.8 units. It floats on water, having a density of 966 kg/m^3.
Plastic *Sulphur* is soft and can be pulled into transparent threads. Its density is 1960 kg/m^3 and it does not conduct electricity (electrical conductivity 10^{-21} units).
Tin is a bright white metal with a density of 7285 kg/m^3 at ordinary temperatures. It did not tarnish in France so the buttons on the uniforms of soldiers in Napoleon's army were made from this metal. During the winter of 1812 Napoleon's army was in Russia. The buttons turned to a dull grey powder and dropped off. Its electrical conductivity is 8.7 units.

Use some of this information to discuss the four patterns. Remember that in classifying you have to decide what information is important and what is not relevant.

(b) Triads
J.W. Dobereiner found groups of elements with similar properties. For example:

Element	Relative atomic mass
Li	6.9
Na	23.0
K	39.1

The relative atomic mass of sodium is 23. It is the average of the relative atomic masses of lithium and potassium. This pattern is called the Law of Triads.

   Use the Law of Triads to estimate the missing information in the tables below.

(i)

Element	Relative atomic mass
Cl	35.5
Br	–
I	126.9

(1)

(ii)

Element	Relative atomic mass
S	32
Se	79
Te	–

(1)

(iii)

Element	Relative atomic mass
Ca	–
Sr	87.6
Ba	137.3

(1)

(c) Octaves

In 1866 John Newlands presented a paper on his Law of Octaves. He listed the known elements in order of their relative atomic masses. He claimed that chemical properties were repeated every eighth element.

1. H   2. Li   3. Be   4. B   5. C   6. N   7. O
8. F   9. Na   10. Mg   11. Al   12. Si   13. P   14. S
15. Cl   16. K   17. Ca   18. Cr   19. Ti   20. Mn   21. Fe

(i)   Find two triads (groups of similar elements) in these columns. *(2)*
(ii) There were only 66 known elements in 1866. How did this affect the pattern? *(1)*
(iii) Find two elements which are badly placed in the columns above and explain how their position is unsatisfactory. *(4)*

(d) Periodicity of physical properties

In 1869 Lothar Meyer plotted graphs of physical properties against relative atomic mass.

Element	Relative atomic mass	Density (kg/m^3)	Relative atomic mass ÷ density
Li	7	533	
Be	9	1846	
B	11	2466	
C	12	2266	
Na	23	966	
Mg	24.3	1738	
Al	27	2698	
Si	28	2329	
P	31	1820	
S	32	2086	
Cl	35.5	2030 (at −160 °C)	
K	39	862	
Ca	40	1530	
Cr	52	7194	
Ni	58.7	8907	
Zn	65.4	7135	
Se	79	4808	
Rb	85.5	1533	
Sr	87.6	2583	

(i)   Copy the table and calculate the data needed for the fourth column. *(4)*
(ii) Plot a graph of the data in the fourth column against relative atomic mass. *(4)*
(iii) List the elements which appear at the peaks of the graph. *(2)*
(iv) Mark on the graph a circle to show the expected position of fluorine (relative atomic mass 19). Work out its expected density in the solid state. *(3)*
(v)   Note the position of chlorine. Mark with a circle the expected position of bromine. *(1)*

(vi) In the 1860s Li was the 2nd element, Na was the 9th, K the 16th and Rb the 30th. Explain whether your graph supports Newlands' Law of Octaves. *(3)*

(e) A periodic table which works

Also in 1869, Dmitri Mendeleev produced a Periodic Table. Like Newlands, he arranged elements in order of their relative atomic masses. But he left gaps, assuming that some elements were still undiscovered. This gave vertical groups of elements with much more similar chemical properties than those suggested by Newlands.

In Group III (which we now call Group IIIb) he had
Boron
Aluminium RAM 27, density 2698 kg/m^3, oxide Al$_2$O$_9$, melting point 660 °C, boiling point 2520 °C.
'Eka-aluminium' (an element undiscovered in 1869).
Indium (discovered in 1863) RAM 115, density 7290 kg/m^3, oxide In$_2$O$_3$, melting point 157 °C, boiling point 2050 °C.

Mendeleev predicted that an element 'eka-aluminium' would be found which had properties which were the average of those of aluminium and indium.

In 1875 gallium was discovered. It had the following properties:
RAM 70, density 5905 kg/m^3, oxide Ga$_2$O$_3$, melting point 30 °C, boiling point 2200 °C. How well did gallium fit the predicted properties of 'eka-aluminium'? *(10)*

(f) Trying more patterns

In 1914 Rydberg suggested a pattern for the number of elements in pairs of periods in the Periodic Table. The two short periods (2 and 3) from Li to Ne and Na to Ar contain altogether $2 \times 8 = 4^2$ elements. The two long periods (4 and 5) from K to Kr and Rb to Xe contain $2 \times 18 = 6^2$ elements.

(i)   If the pattern still works, these should be followed by two periods (6 and 7) with a total of $2 \times 32 = 8^2$ elements. Examine a copy of the modern Periodic Table and explain whether or not this is a reasonable idea. *(2)*
(ii) Following this pattern, the two short periods should have above them periods containing 2^2 elements of which hydrogen and helium are known. Rydberg suggested that these come between hydrogen and helium: nebulium (RAM 1.31) and coronium (RAM 2.1). Both were thought to have been

observed by astronomers in the light from nebulae. Bright green lines seen in the spectrum of light from the sun during the eclipse of 1869 were thought to show coronium. Nasini, Anderlini and Salvadori claimed to have found it in volcanic gases in 1893. Look up in a textbook 'atomic number', 'proton' and 'structure of the atom' and discuss the existence of nebulium and coronium. (6)

(g) **X-ray inspection**
In 1913 Moseley fired electrons at elements or their compounds and measured the frequency of the X-rays produced. The diagram shows part of the graph he obtained by plotting $\sqrt{\text{(frequency)}}$ against Z, the 'atomic number' of the element. At that time 'atomic number' was simply a number which gave the order of the element in the Periodic Table.

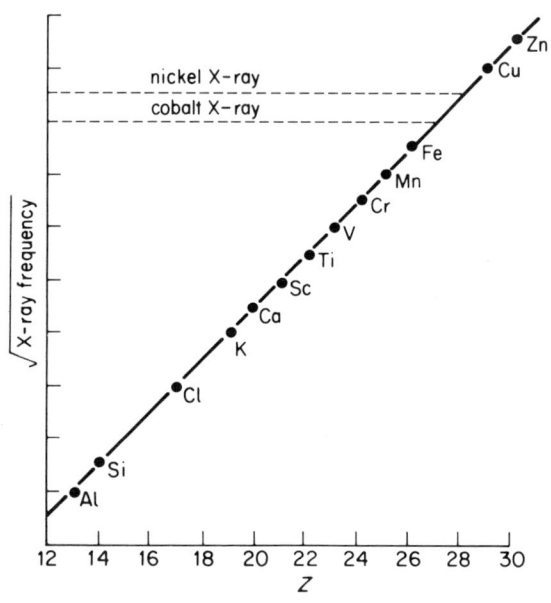

(i) What problems do you find when you try to fit cobalt and nickel into the pattern? (2)

Element	Relative atomic mass
Iron	55.85
Cobalt	58.93
Nickel	58.71
Copper	63.55

(ii) What is the problem with argon? (2)

Element	Relative atomic mass
Chlorine	35.45
Potassium	39.10
Argon	39.95

(iii) Look up in a textbook 'isotopes' and try to explain why there are these problems. (2)
(iv) Would Rydberg's 2^2 idea (part (f)) fit in with Moseley's information about Z? (3)

(h) **GCSE Science Criteria**
Here are some of the GCSE Science Criteria:
– Recognise patterns in data.
– Form hypotheses (make intelligent guesses).
– Deduce relationships.
In parts (a) – (g) you have seen that this is how Chemistry knowledge is obtained.
–Make decisions based on the examination of evidence and arguments.
This is what you have been doing in earlier parts of this question.
–Recognise that the study and practice of science are subject to various limitations and uncertainties.
This really takes you into the next chapter: Evaluation. But try writing a short essay on 'limitations and uncertainties of Chemistry' using ideas from this question to help you to explain what you mean. (10)

4 Evaluation

1

The table shows some properties of four substances.

	Substance			
	P	Q	R	S
Melting point (°C)	318	800	−114	−117
Boiling point (°C)	1375	1410	−85	78
pH of its solution	14	7	1	7

Pupils were asked to write down one pattern shown by the substances. Most of the class gave one of the following patterns. Which one is correct?

A. The higher the boiling point, the more alkaline the solution.
B. The higher the melting point, the more alkaline the solution.
C. If it is a liquid at room temperature, it has a neutral solution.
D. If it is a solid at room temperature, it has an alkaline solution.

2

Jo is proud of the good tea he makes. He always follows the same routine.

First he puts tea leaves in the teapot. Then he adds boiling water. He lets the pot stand for a few minutes. He pours the tea into a cup, using a strainer to catch the tea leaves. Then he puts milk and sugar in the cup and stirs the mixture.

Which of these is the correct order of the scientific names for the way Jo makes a cup of tea?

A. Dissolving, filtering, evaporation.
B. Dissolving, filtering, dissolving.
C. Evaporating, dissolving, filtering.
D. Filtering, evaporating, filtering.

3

At school Dawn was caught wearing nail polish. Her teacher sent her to take it off. Dawn found that washing her hands in hot water did not take off the polish, so she went to the science technician for help. He gave her some propanone (acetone) and cotton wool. Cotton wool soaked in propanone cleaned the polish off her nails but took away some of their natural oil and left them feeling dry.

This story shows that:
A. nail polish is soluble in water.
B. nail polish is soluble in propanone.
C. cotton wool scrapes off nail polish.
D. dry nails do not hold polish.

4

The graphs show the percentage of energy used in the UK which came from coal, oil and natural gas.

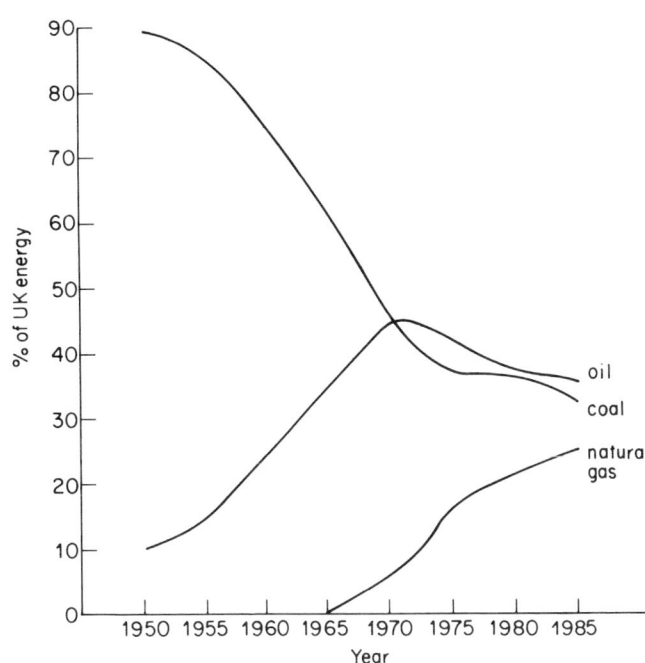

Which of the following statements is correct?
A. The percentage of energy that comes from oil is still increasing.
B. Coal is no longer used for energy.
C. As the percentage of oil used has increased, the percentage of coal used has decreased.
D. The percentage of energy that comes from natural gas is still increasing.

5

Barium compounds are poisonous, but to X-ray the digestive system, patients are given a 'barium meal' of barium sulphate. This makes the digestive system absorb X-rays so that it stands out as white tubes on X-ray photographs (negatives). Which of the following facts best explains why patients are not poisoned when they swallow a 'barium meal'?
A. X-rays produce electrons when they are absorbed by barium sulphate.
B. Barium sulphate is not soluble.
C. The barium sulphate has a higher relative molecular mass than most other barium salts.
D. Any undigested barium sulphate is passed out of the body in faeces only a few hours after the 'barium meal' is swallowed.

6

Oil spills from tankers at sea can cause a great deal of pollution damage. Clean-up operations are expensive and tanker owners have to pay compensation.
(a) Bar chart 1 shows how the number of spills of more than 5000 barrels has changed between 1975 and 1986.

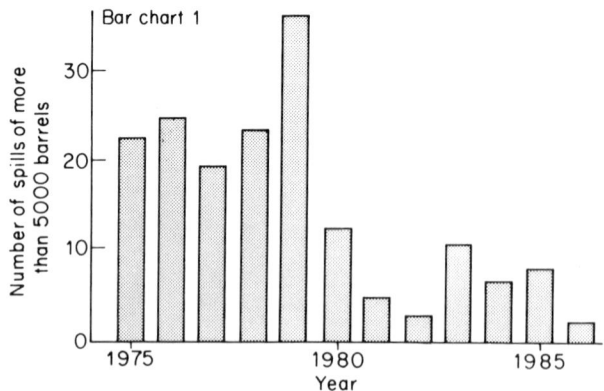

Consider the information in bar chart 1 and decide which of the following statements evaluates it correctly.
A. Less oil was spilt in 1986 than in 1985.
B. There were fewer spills of more than 5000 barrels in 1986 than in any year since 1975.
C. More oil was spilt in 1979 than in any other year between 1975 and 1986.
D. In 1985 there were fewer tankers carrying oil than in 1975.

(b) Since 1973 compensation for pollution damage has been paid by the CRISTAL system (Contract Regarding an Interim Supplement to Tankers' Liability for oil pollution).

Bar chart 2 shows the *cumulative* compensation paid by CRISTAL. On average, the compensation money was paid out one or two years after the accident. By 1986 a total of 58 million US dollars had been paid out.

Consider the information in bar chart 2 and decide which of the following statements evaluates it correctly.

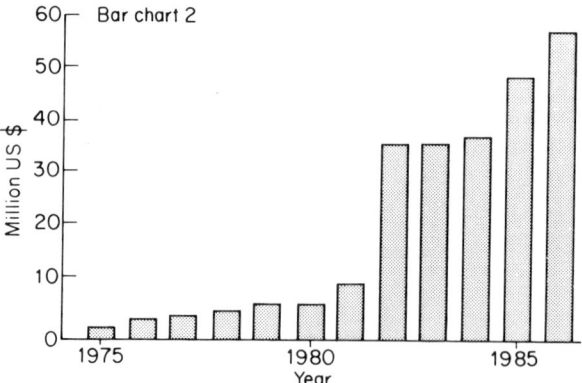

A. More money was paid out in 1982 than in any other year between 1975 and 1986.
B. The amount of money paid out each year has always been greater than that paid out the previous year.
C. The amount of money paid out during 1983 was the same as during 1982.
D. Regulations changed in 1982 so that much more money had to be paid out for the same amount of damage.

(c) Now consider both bar charts and decide which of the following statements is NOT a reasonable evaluation of the data.
A. Tanker operators became careful after the 1979 spills led to the payment of big claims in 1982.
B. On average, the compensation paid per accident has increased between 1975 and 1986.

C. If the trend in bar chart 1 continues, bar chart 2 will flatten off.

D. If the trend in bar chart 1 continues, bar chart 2 will start to show a downward trend.

7

The 'Bonsoir' dairy firm produces cheese and yoghurts. The diagram shows two of their popular products.

 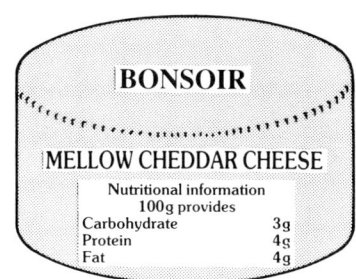

BONSOIR YOGURT

Nutritional information
100g provides
Carbohydrate   4g
Protein   4g
Fat   0.1g

BONSOIR

MELLOW CHEDDAR CHEESE

Nutritional information
100g provides
Carbohydrate   3g
Protein   4g
Fat   4g

(a) Draw a bar chart to show the figures for the nutritional information given on each product. *(4)*

(b) Explain which of these two Bonsoir products would be most suitable for someone trying to lose weight. *(2)*

(c) For each of the components of the yoghurt and cheese, explain how they are used by the human body. *(3)*

(d) Bonsoir have developed a new brand of cheese which has half the normal amount of fat. They advertise this as 'very low fat cheese'. Comment on the truthfulness of their advertisement. *(3)*

8

Ultraviolet light from the sun degrades the surface of the plastic PVC (polyvinyl chloride or poly(chloroethene)). It changes it to a polyenyl structure. This surface layer is a good absorber of ultraviolet light and protects the PVC beneath the surface. Window glass absorbs most ultraviolet light.
Explain the following:

(a) Thin PVC film used for packaging works well in the shop and in the kitchen. If it is left as litter in the street, it is broken down by the sun. *(3)*

(b) Coloured PVC film is not degraded as quickly as clear PVC film when left in the sun. *(1)*

(c) Sheets of PVC, 5 mm in thickness, can be used for many years as roofing material. *(2)*

*9

The Hotel Grand cleans its linen table cloths by boiling them with bleach. Most of the stains are removed, but some of the protein stains like egg and gravy are fixed into the linen by the hot water. The table cloths also begin to develop holes after being boiled in bleach a few times.

A new manager joins the staff and decides to make the cloths wash cleaner and last longer.

(a) She tells staff that as the bleach reacts to remove stains, it forms hydrochloric acid. This is what makes holes in the linen. Rinsing the cloths in plenty of clean water will help; so will putting washing soda (sodium carbonate) in the final rinse.

  (i) Explain how rinsing in plenty of clean water should help to prevent holes. *(2)*

  (ii) Suggest how washing soda will stop the hydrochloric acid making holes in the linen cloths. Use a word equation to help you. *(4)*

(b) As new table cloths are bought, they are washed in a 'biological' washing powder. This contains enzymes that 'digest' the protein stains. Not only are the cloths clean, but the powder works at a water temperature lower than 37 °C.

  (i) Suggest why the washing water must stay below 37 °C if the biological powder is to work. *(2)*

  (ii) Explain two advantages for the Hotel Grand if it does all its washing using the biological powder. *(4)*

  (iii) Suggest one disadvantage the hotel may find when using this powder. *(1)*

10

Soap and detergents for washing clothes have molecules with one end which 'likes' water (hydrophilic). The other tries to escape water (hydrophobic) by burying into particles of grease, dirt and cloth. These molecules can be drawn like this:

Diagram 1

hydrophobic   hydrophilic

The hydrophilic ends push against one another by electrostatic repulsion. So the grease and dirt get pushed off the cloth.

Diagram 2

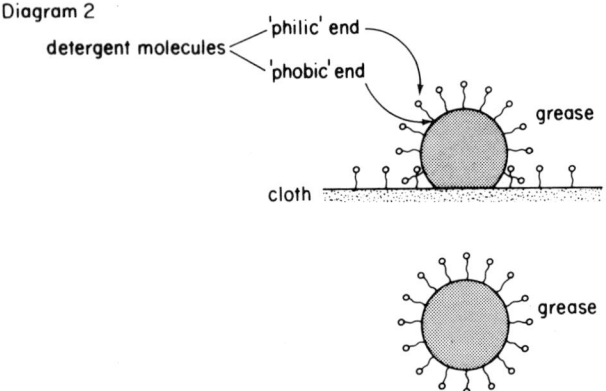

Hair conditioners that fluff out hair and 'give it body' contain molecules that behave like soap molecules. These conditioner molecules attach to each hair like this:

Diagram 3

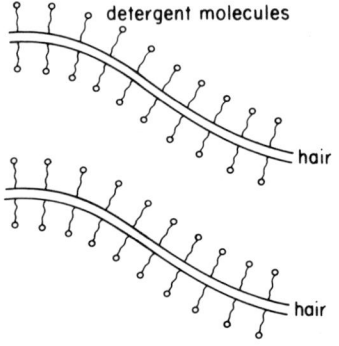

(a) Suggest an explanation for the way conditioner molecules can give hair 'body'. (2)
(b) The instructions on a bottle of conditioner say 'First shampoo hair, rinse and towel dry. Massage conditioner evenly through the hair and rinse thoroughly'.
    Suggest reasons for the following instructions:
    (i)   the first 'rinse'. (1)
    (ii) 'towel dry'. (1)
    (iii) 'massage evenly'. (1)
    (iv) 'rinse thoroughly'. (1)

Many shops sell water filters like the one in the diagram. You can use a filter jug like this to purify your drinking water.

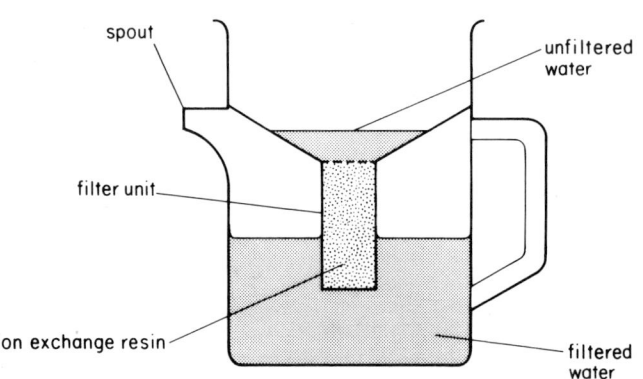

(a) Suggest and explain two ways in which the water is purified. (4)
(b) The makers say the jug will purify up to 40 litres of water and then the 'filter unit' needs replacing.
    (i) Explain why the unit will need to be replaced. (2)
    (ii) Some users could find that the unit purifies more than 40 litres of water. Explain why. (1)
(c) The makers say that the filtered water will make better tasting tea and coffee. Why is it difficult to prove this statement scientifically? (2)

*12 Wine making

The Smith family enjoy making home-made wines. One of their favourites is dandelion wine made by this recipe:

Ingredients
4.5 dm³ (l) dandelion petals
1.3 kg sugar
4.5 dm³ (l) boiling water
1 orange
1 lemon
20 g wine making yeast
 5 g yeast nutrient

Method
1. Wash the dandelion flowers and then cover them with boiling water. Leave to stand for three days. Strain the flowers out of the liquid.
2. Bring the liquid to the boil. Pour the liquid over the orange and lemon rind and sugar. Stir until all the sugar is dissolved.
3. Allow to cool to room temperature. Add juice of the lemon and orange.
4. Add the yeast and the yeast nutrient. Stopper the demijohn as shown in diagram 1. Leave in a warm place to ferment.

Diagram 1

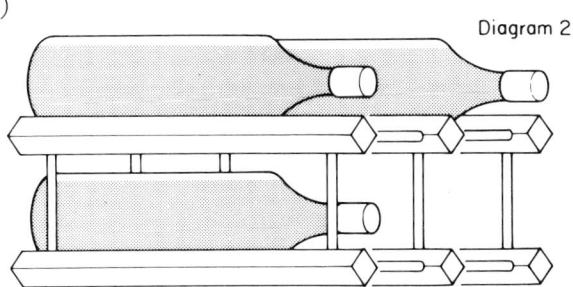

demijohn

fermenting liquid

water

airlock

(a) Why are the dandelion flowers washed and then covered with boiling water? (2)

(b) (i) Give a reason why hot liquid rather than cold liquid is added to the sugar (in step 2). (2)

(ii) Why is the liquid cooled to room temperature before the yeast is added? (1)

In making wine the following reactions occur:

$$sugar + water \xrightarrow{acid} glucose + fructose \ [1]$$

$$glucose + fructose \xrightarrow{yeast} ethanol + carbon \ dioxide \ [2]$$

(c) (i) What is the source of the acid in reaction [1]? (1)

(ii) How could the apparatus shown in diagram 1 be modified to prove that carbon dioxide is given off? What would you observe? (3)

(iii) The dandelion flowers do not take part in the reactions. Why are they added? (1)

(iv) Why is yeast nutrient included in the ingredients? (2)

(d) (i) Why is the air lock attached to the demijohn instead of using a rubber bung? (3)

(ii) Explain why the demijohn would not be used half-full. (2)

(iii) There is always a small space left at the top of the demijohn. Why is it not filled completely? (1)

(e) (i) Before the wine is bottled all solid matter must be removed. What process could be used to remove the solids? (1)

(ii) Explain why the bottles used must be sterilised and dried before the wine is poured into them. (2)

(iii)

Diagram 2

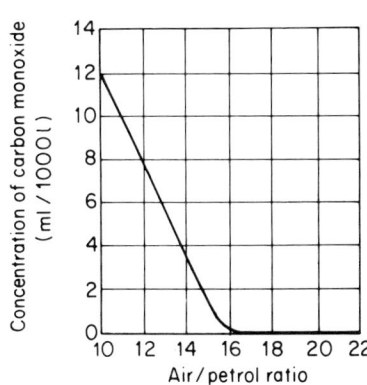

Bottles of wine are stored on their sides rather than upright.

Give a possible reason why this method of storage is used. (2)

*13 Pollution by exhaust fumes

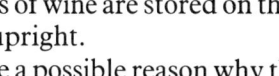

There are now laws to reduce pollution from motor vehicles. In 1968, motor vehicles in the USA produced

59% of the carbon monoxide in the air
49% of the hydrocarbon in the air
35% of the nitrogen oxides in the air
3% of the solid particles in the air
1% of the sulphur dioxide in the air.

(a) Which exhaust gases do you think it most important to control? (3)

(b) Carbon monoxide is produced when fuel is not burned completely. Complete burning produces carbon dioxide. Nitrogen oxides are produced when air (containing nitrogen and oxygen) is at a very high temperature. At one time Ford and other companies tried to reduce pollution by feeding air into the exhaust gases as they came out of the engine cylinder. This is where the gases are at the highest temperature.

There was less of two polluting gases but more of another. Which gases decreased and which increased? Explain your answer. (5)

(c) Some engines are fitted with catalytic converters. These contain platinum to increase oxidation. How does this improve the exhaust gases? (2)

(d) In some countries, cheap alcohol made from sugar cane is added to petrol. This reduces the amount of sulphur dioxide in exhaust gases. Suggest why. (2)

(e) Carbon monoxide is a poisonous gas. The graph shows how the carbon monoxide in car exhausts depends on the air/petrol ratio.

(i) A petrol engine normally works with an air to petrol ratio of 15:5. What concentration of carbon monoxide does this produce? *(1)*

(ii) Explain why it is dangerous to run a car engine in a closed garage. *(2)*

(iii) To start a car engine there must be an air to petrol mixture which is rich in petrol. A rich mixture is obtained by using the choke control. It is more dangerous to leave a car engine running in a closed garage before a journey than at the end of a journey. Explain why. *(2)*

14 Fizzy drinks

The Phizzy company have launched a new range of drinks, called Phizz-a-pac. The Phizz-a-pac range are all fizzy drinks and are packed in plastic bottles which are designed to be ideal for people taking packed lunches to school or work.

The diagram shows one of the Phizz-a-pac bottles. The 'fizz' in the drink which it holds is caused by carbon dioxide gas. The company dissolves carbon dioxide in the drink. As the diagram shows, the gas stays in solution until the bottle cap is unscrewed.

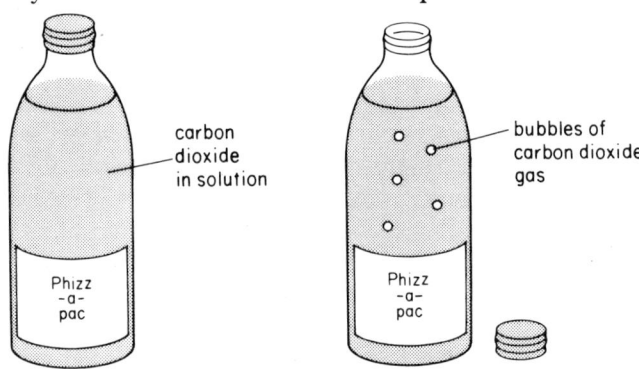

carbon dioxide in solution

bubbles of carbon dioxide gas

Phizz -a- pac

Phizz -a- pac

Carbon dioxide dissolves in water in two stages which can be summarised by the following equations:

$$CO_2(g) \rightarrow CO_2(aq) \qquad [1]$$

$$CO_2(aq) + H_2O(l) \rightarrow H_2CO_3(aq) \qquad [2]$$

(a) (i) Use equations [1] and [2] to write an equation or equations to describe what is happening in the open Phizz-a-pac bottle. *(2)*

(ii) State one assumption you have had to make in answering part (i). *(1)*

(b) Explain the following:

(i) Carbon dioxide stays in solution when a Phizz-a-pac bottle is closed. *(2)*

(ii) If the bottle top is not screwed on tight, the

Phizz-a-pac it contains will lose its carbon dioxide and go 'flat'. *(2)*

(c) Explain which of the bottles in the diagram you would expect to contain Phizz-a-pac with the lowest pH. *(3)*

(d) Suggest and explain two design features that you would expect Phizz-a-pac bottles to have. *(2)*

*15 Recycling plastics

Britain produces 30 million tonnes of refuse each year.

(a) Different types of refuse can be separated by air classification. If air is travelling fast enough, some materials will be carried with it. The accompanying diagrams show the air speeds needed to carry certain materials.

What happens if the air speed is 10 m/s? *(2)*

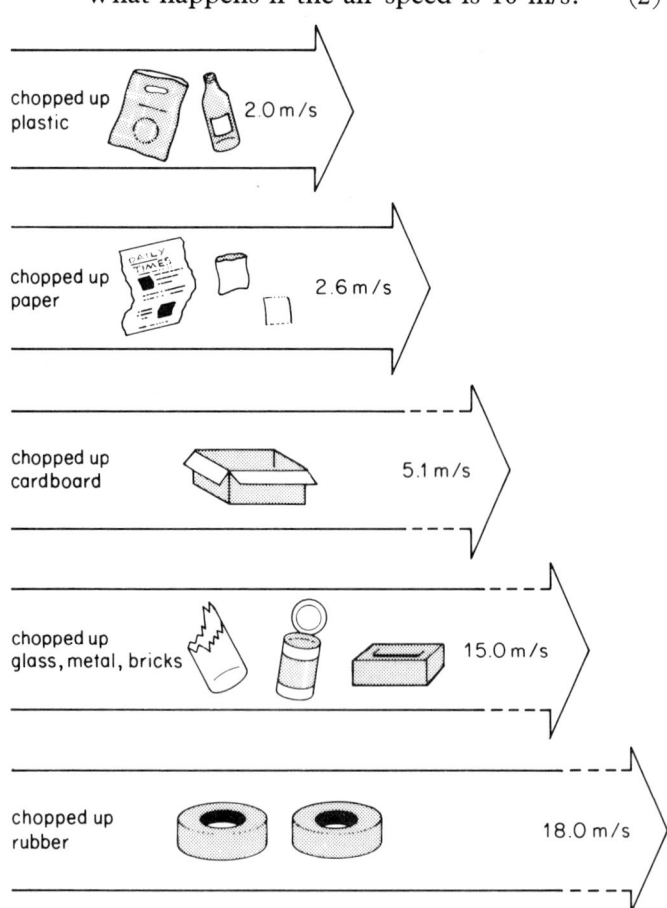

chopped up plastic   2.0 m/s

chopped up paper   2.6 m/s

chopped up cardboard   5.1 m/s

chopped up glass, metal, bricks   15.0 m/s

chopped up rubber   18.0 m/s

(b) (i) Give two reasons why paper and plastics are usually burned together in energy recycling schemes. *(3)*

(ii) Explain carefully three ways in which this energy could be used. Bear in mind that refuse tips are often some distance from towns and factories. *(6)*

(c) The table shows how much paper and plastic is contained in an average tonne of refuse, together with the calorific values of these materials.

Material	Mass in 1 tonne of refuse (kg)	Calorific value (MJ/kg)
Paper	50	2.8
Plastic	430	7.3

(i) If the plastic in a tonne of refuse is burned, how much energy is released? *(1)*

(ii) If all the plastic in Britian's refuse could be burned, how much energy could be released? *(1)*

(iii) Suggest why it is usually difficult to get the paper in refuse to burn. *(1)*

(d) (i) Polyethylene (polyethene), polypropylene (polypropene) and polystyrene (polyphenylethene) contain only hydrogen and carbon. What gases are likely to be produced when they are burned? *(1)*

(ii) PVC (polyvinyl chloride or polychloroethene) contains chlorine. What acid gas is likely to be produced when it is burned? *(1)*

(iii) Dust often sticks to paper and plastic in a refuse tip. If some is limestone dust, the quantity of acid gas produced by burning PVC is less than expected. Explain this. *(2)*

(e) A Japanese company operates the following process on plastics waste:
Crushed and melted polymer is put into a cracking reactor and the polymer is pyrolysed into vapour at a temperature of 700 °C. The vapour is passed through a condenser where oil is separated from gas. 90% of the original plastic waste is recovered as oil. It has a flash point of 30 °C (it catches fire at 30 °C). It has a calorific value of 45 MJ/kg. The cost of building a plant to process 36 tonnes of plastics waste per day was £150 000 in 1971 when the income from the sale of the oil produced was £97 000 per year.

(i) Explain the terms:
1. cracking. *(1)*
2. pyrolysed. *(1)*

(ii) How much oil is produced when 36 tonnes of plastics waste is processed? *(1)*

(iii) 1. Besides the cost of building the plant, what other costs are there in operating this process? *(3)*

2. Explain whether it would be worth building another plant at a time when building cost has increased by four times and oil is twice the price. *(2)*

16 Salt works

(a) The solubilities of sodium chloride and sodium sulphate were measured at various temperatures. The values are given below.

Temperature (°C)	Sodium chloride (g per 100 g water)	Sodium sulphate (g per 100 g water)
0	35.7	4.7
10	35.8	9.1
20	36.0	20.4
30	36.2	41.0
40	36.5	48.2
50	36.8	46.7
60	37.2	45.2
70	37.6	44.1
80	38.1	43.3
90	38.6	42.7
100	39.2	42.3

Describe the pattern linking solubility to temperature for
(i) sodium chloride. *(2)*
(ii) sodium sulphate. *(3)*

(b) The Rightwich Salt Company extracts rock salt from the local rocks. Rock salt is mainly sodium chloride.

Diagram 1

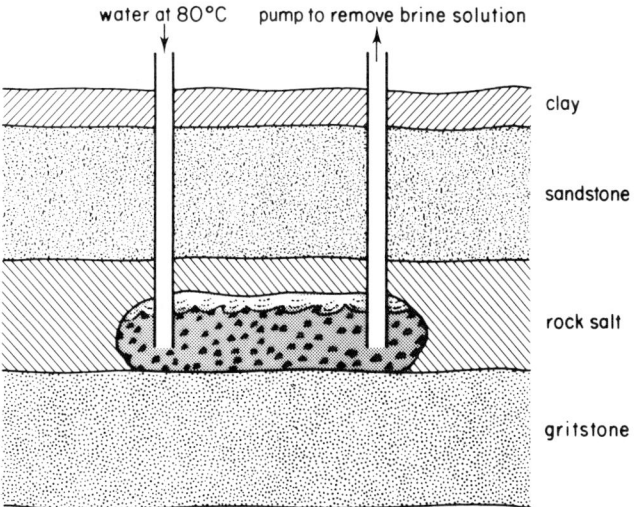

Water (at 80 °C) is pumped down to dissolve the rock salt. The rock salt solution, called brine, is pumped up another pipe.
Give a reason why water is used at 80 °C instead of 20 °C. *(2)*

(c) The rock salt contains about 10% sodium sulphate. Using the information given in part (a), suggest how the sodium sulphate could be separated from the brine. *(3)*

(d) The Rightwich Salt Company adds barium chloride solution to the brine. The following reaction occurs:

$$BaCl_2 + Na_2SO_4 \rightarrow BaSO_4 + 2NaCl$$

The solubility of barium sulphate at 20 °C is 0.0002 g per 100 g of water.
  (i) Explain one advantage of this method compared to the one you suggested in part (c). (2)
  (ii) How could the barium sulphate be removed from the brine? (3)

(e) At one time the Rightwich Salt Company collected the sodium chloride from the purified brine by evaporating the solution in large open vats. Now they get the sodium chloride by evaporating the brine at 40 °C. The pressure above the brine is slowly reduced.

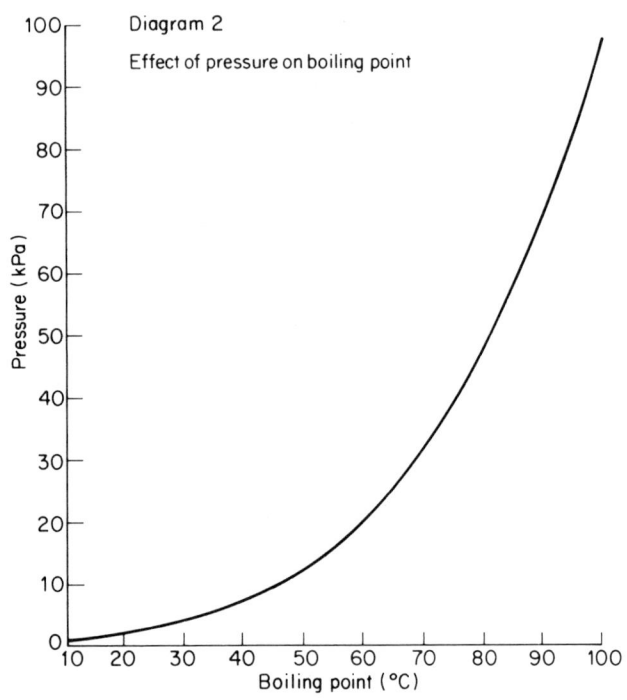

Diagram 2

Effect of pressure on boiling point

Give the advantages and disadvantages of the modern method compared with the old method. (5)

17 Refining uranium

(a) Uranium is slightly radioactive. As it decays, radium is produced which is also radioactive. Eventually a non-radioactive isotope of lead is formed.

The black mineral pitchblende is the commonest ore of uranium. It contains uranium dioxide. It also contains compounds of radium and lead.

When it is mined the ore contains sandy material as well as the mineral. If the ore contains as much as 0.15% uranium dioxide, it is usually worth mining. The ore is found in Australia, South Africa, the USA and other countries. It is usually processed to a concentration of at least 60% uranium before being shipped to Britian where it is refined.
  (i) Why are radium and lead compounds always found in the same ore as uranium? (1)
  (ii) Why would ore containing less than 0.15% uranium dioxide not be worth mining? (1)
  (iii) Why is its concentration increased before it is shipped to Britain? (2)

(b) The diagram on the following page shows how pure uranyl nitrate is obtained from concentrated uranium ore.

  (i) In tank A the uranium oxide in the ore dissolves in nitric acid to form uranyl nitrate. Suggest two reasons why the tank is heated by steam. (2)
  (ii) All nitrates are soluble in water. Suggest two materials which could be caught by the filter. (2)
  (iii) In tank C_1 the nitric acid containing dissolved nitrate is mixed with an organic solvent. Look at the diagram of tank C_2.
    1. Why does the organic solvent float on top of the nitric acid so that they form two separate layers? (2)
    2. Suggest why there are impurities left in the nitric acid and only the uranyl nitrate passes from the acid into the organic solvent. (2)
  (iv) In tank D_2 the uranyl nitrate has passed from the organic solvent into water.
    Tank C_2 is cool. Tank D_2 is hot.
    What does the diagram tell you about the solubility of uranyl nitrate in the organic solvent and in water or water-based solutions? (4)
  (v) The organic solvent does not get used up. Why is this? (1)
  (vi) What has been achieved by tanks C and D? (2)
  (vii) What are the pieces of equipment labelled X? (1)
  (viii) Why does steam go into tank E at the top and out at the bottom? (1)
  (ix) What gas escapes through opening O? (1)

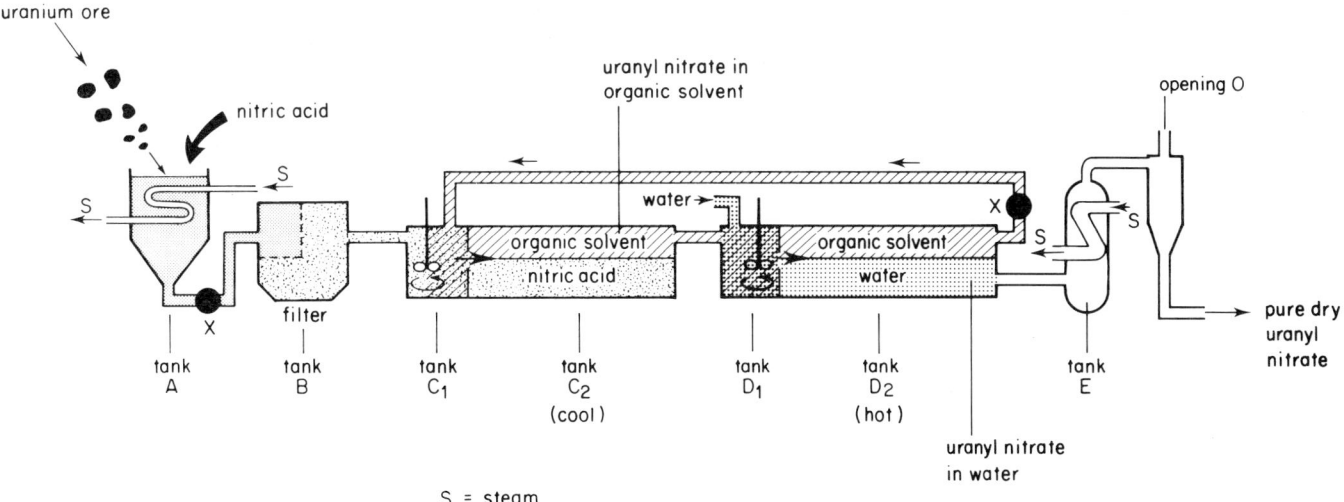

uranium ore

nitric acid

uranyl nitrate in organic solvent

opening O

S

S

water →

organic solvent
nitric acid

X

organic solvent
water

X

S
S

pure dry uranyl nitrate

filter

X

tank A

tank B

tank C₁

tank C₂
(cool)

tank D₁

tank D₂
(hot)

tank E

uranyl nitrate in water

S = steam

---

*18 Degrading plastics

Some plastics are biodegradable.

At present, polyethenes up to a relative molecular mass of 4800 can be digested by bacteria. Polyethenes have the structural formula

$$\left(-\underset{\underset{\text{H}}{|}}{\overset{\overset{\text{H}}{|}}{\text{C}}} - \underset{\underset{\text{H}}{|}}{\overset{\overset{\text{H}}{|}}{\text{C}}} - \right)_n$$

Chlorinated lower alkanes (such as monochlorobutane, C_4H_9Cl) can be used by bacteria as a source of carbon. Diagram 1 shows how the rate of bacteria growth depends on the percentage of chlorine in the compound. It is based on research published in Germany in 1963 (A. Schwartz).

Diagram 1

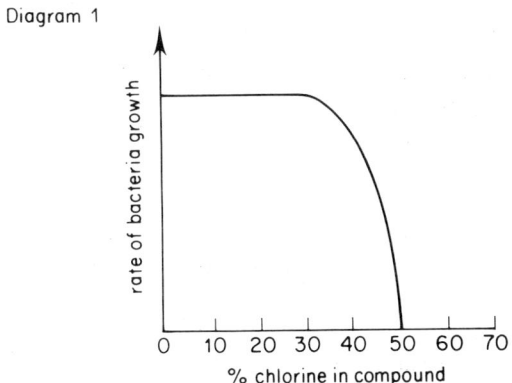

Polychloroethene (PVC), has the structural formula:

$$\left(-\underset{\underset{\text{H}}{|}}{\overset{\overset{\text{H}}{|}}{\text{C}}} - \underset{\underset{\text{Cl}}{|}}{\overset{\overset{\text{H}}{|}}{\text{C}}} - \right)_n$$

(Relative atomic masses: $(C = 12; H = 1; Cl = 35.5)$

Some polymers containing unsaturated bonds are biodegradable.

$$\overset{}{>}C=C\overset{}{<}$$

(a) Evaluate the statement: 'at present polyethenes up to a molecular weight of 4800 can be digested by bacteria'.
In what way are these facts likely to be affected by evolution? (2)

(b) Polyethene is an alkane chain but its name suggests an alkene. Explain this. (2)

(c) Natural rubber is *cis*-polyisoprene (poly-2-methylbut-2-ene). It has the structural formula:

$$(-CH_2-\underset{\underset{CH_3}{|}}{C}=CH-CH_2-)_n$$

Older electrical wiring in buildings used cables insulated with rubber and cased in lead.

Diagram 2

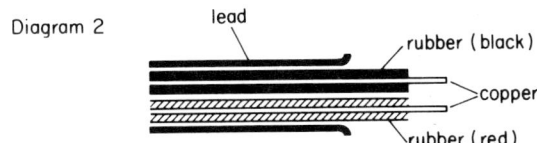

lead

rubber (black)

copper

rubber (red)

41

(i) Explain why such wiring is now being replaced. (2)

(ii) Which sections of such wiring are likely to be causing problems? (1)

(d) More than twenty years ago, wiring in buildings started to be replaced by 'PVC/PVC' cable.

Diagram 3

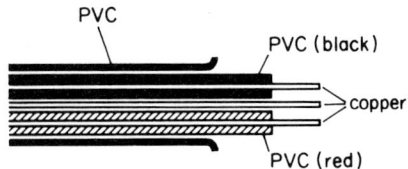

Explain why some wiring engineers believe this will last for ever. (4)

19 Radon gas

Radon is a gas. The isotope Rn-222 (relative atomic mass 222) emits alpha particles and has a half-life of 3.82 days. Levels are high in some areas of radioactive rock such as granite. It is a particular problem in Cornwall where the granite is cracked, allowing the gas to escape.

(a) Explain why there is less of a problem with radon in
(i) London which is on clay. (1)
(ii) Aberdeen which is on solid granite. (1)

(b) During the war everyone in Britain was issued with a gas mask.

Diagram 1

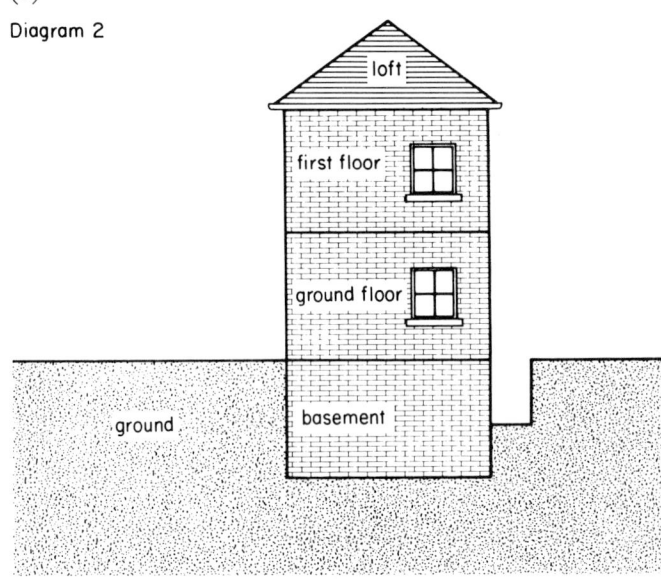

carbon absorbs big molecules

clean air

air containing big molecules of poisonous substances

This was in case poisonous gas was used. The mask contained carbon. Carbon absorbs molecules of poisonous substances. Cigarette smoke is made up of small particles of carbon. Explain why the lung cancer risk is higher for people who breathe in cigarette smoke in high radon areas. (4)

(c)

Diagram 2

loft

first floor

ground floor

ground

basement

Give two reasons why radon-222 is most likely to be found in the basement of this house. (2)

(d) Radon can be emitted by building materials such as concrete blocks. This is a particular problem in Sweden and the USA. Why does the gas come from some building materials? (2)

(e) In Britain the National Radiological Protection Board estimates that there may be up to 100 000 houses which give their occupants a radiation dose of 5 millisieverts per year. (1 mSv/yr is the maximum permissible dose to the public from the nuclear industry.) In June 1986 a letter appeared in a national newpaper (the Scotsman) making the following claim:

'It has been estimated that draught proofing leads to an extra 100 cancer deaths per year per 100 megawatts saved, due to greater concentrations of natural radioactivity retained within the building.'

A magazine article later called this 'an outrageous assertion'. There are 20 million homes in Britain. A room can be nicely heated by a 1 kW heater.

(i) How much heat loss do you think you could stop from your house by good draught proofing? How can you estimate this? (2)

(ii) How many houses are seriously affected by radon? (1)

(iii) Evaluate the statement made in the letter which appeared in the Scotsman. (3)

20 Ozone

Absorption of radiation from the sun by gases in the earth's atmosphere

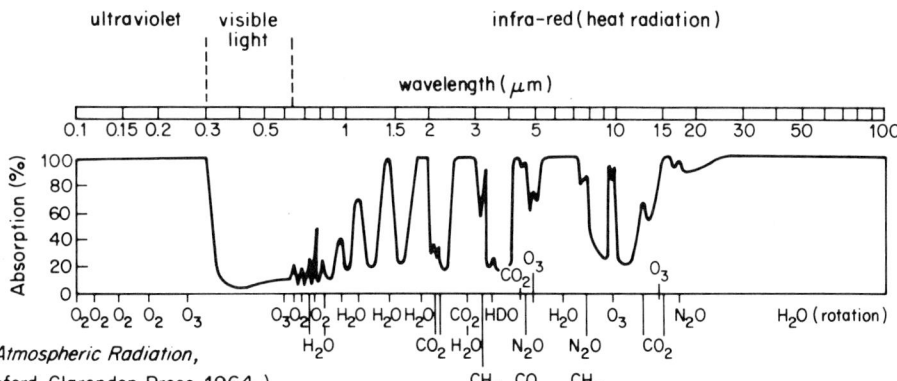

(Source: Goody, R.M. *Atmospheric Radiation,*
I Theoretical Basis. Oxford, Clarendon Press. 1964.)

There is a lot of ozone, O_3, in the upper atmosphere. This 'ozone layer' absorbs some of the radiation from the sun.

(a) (i) The longest wavelength absorbed by ozone is 14.5 μm. Use the graph to find out what other wavelengths are absorbed by ozone. *(4)*

(ii) Ultraviolet radiation can cause skin cancer. How does ozone help to reduce this hazard? *(1)*

b) Long exposure to ozone can cause irritation to the eyes. Concorde flies in the upper atmosphere at a height of 18 kilometres. It is fitted with nickel catalyst crackers to convert ozone to oxygen.

(i) Why is this necessary? *(1)*

(ii) Why is the nickel cracker likely to be of more benefit to air crews than to passengers? *(1)*

(c) When Concorde burns 1 kg of fuel, about 1.3 kg of water vapour and 0.025 kg of nitrogen dioxide are produced.

(i) Concorde uses a hydrocarbon fuel which burns in air. How do you think the nitrogen dioxide is produced? *(1)*

(ii) The following reactions take place in the atmosphere:

1. $O_2 \rightarrow O + O$

$NO_2 \rightarrow NO + O$

These reactions are examples of 'photochemical dissociation' caused by ultraviolet radiation.

2. $O + O_2 \rightarrow O_3$

$O_3 + O \rightarrow 2O_2$

These reactions are due to the presence of free oxygen atoms.

3. (a) $O + H_2O \rightarrow 2OH$ (hydroxide radical)

(b) $OH + O_3 \rightarrow O_2H + O_2$

$O_2H + O \rightarrow OH + O_2$

Overall reaction:

$O_3 + O \rightarrow 2O_2$ hydroxide radical acts as catalyst.

4. $NO + O_3 \rightarrow NO_2 + O_2$

$NO_2 + O \rightarrow NO + O_2$

Overall reaction:

$O_3 + O \rightarrow 2O_2$ nitrogen monoxide, NO, acts as catalyst.

Explain carefully the effects Concorde, flying at between 14 km and 18 km above ground level, could have on the ozone layer. *(4)*

(d) Chlorofluoromethanes are used in aerosol sprays. They dissociate when ultraviolet radiation acts on them. Chlorine atoms are released.

$$CF_2Cl_2 \rightarrow CF_2Cl + Cl$$

(i) Chlorofluoromethanes dissociate in the upper atmosphere but not in the lower atmosphere. Suggest why. *(2)*

(ii) Chlorine atoms damage the upper atmosphere.

$$Cl + O_3 \rightarrow ClO + O_2$$
$$ClO + O \rightarrow Cl + O_2$$

What overall reaction is catalysed by free chlorine atoms? *(1)*

(iii) $ClO + NO_2 \rightarrow ClONO_2$
This compound, chlorine nitrate, is unreactive. Do high-level aircraft increase or reduce the damage done by aerosol sprays? *(2)*

5 Synthesis

*1

The diagram is often called the nitrogen cycle because it summarises the processes which keep nitrogen atoms in circulation.

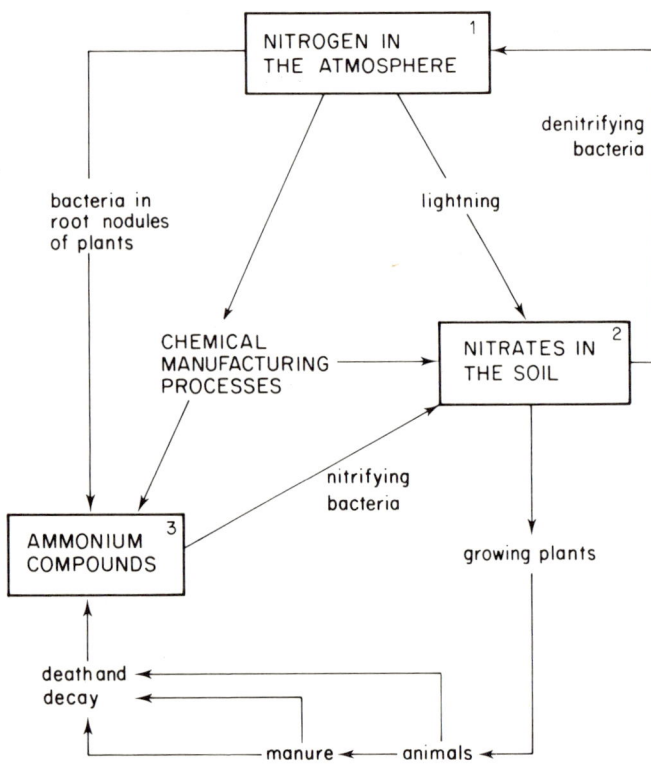

(a) (i) What chemical manufacturing process uses atmospheric nitrogen? (1)
    (ii) Briefly describe this process, using equations to show how the nitrogen is used. (4)

(b) Usually nitrates in the soil (box 2) and ammonium compounds (box 3) get used up at the same rate as they are created. Suggest and explain two ways in which the level of each of them could become unbalanced. (8)

(c) (i) Explain why plants and animals need nitrogen. (2)
    (ii) How do plants obtain their nitrogen? (2)
    (iii) How do animals obtain their nitrogen? (2)

2

(a) Diagram 1 shows the sources of radiation to which people in the United Kingdom are exposed.

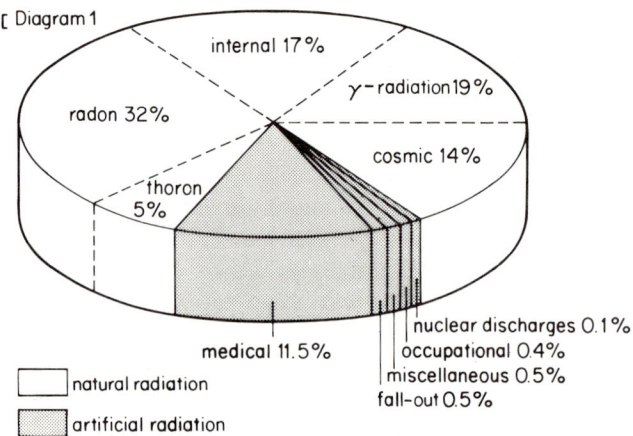

[Diagram 1

- natural radiation
- artificial radiation

(i) What is the total percentage of natural radiation? (1)
(ii) The average radiation dose per year is 2.39 milli-Sieverts (mSv). What annual dose comes from artificial radiation? (3)
(iii) Granite is rock which slowly gives off radon gas. Why is it possible for a family who live in a house built of granite to receive more radiation than a family who live close to a nuclear power station? (4)

(b) Diagram 2 shows the doses of radiation from various sources and the effect they have on human beings.
(i) Why do radiographers (workers who take X-ray photographs) not stay with a patient who is being X-rayed? (2)
(ii) Suggest why radiation is used to treat cancer. What side effects might the patient expect from the treatment? (4)
(iii) Give a possible explanation why a family on holiday for one week in Cornwall is exposed to more radiation than if they had spent a week at Blackpool. (2)
(iv) Give a possible explanation of how passengers on transatlantic flights receive extra radiation. (1)

Diagram 2

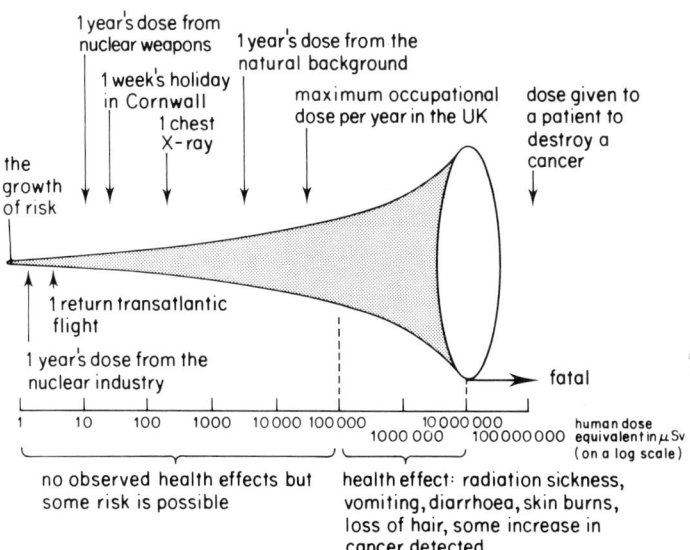

(a) Using the letters X and Y, write down the total energy
   (i) used in breaking bonds. *(1)*
   (ii) released as new bonds form. *(1)*

(b) Is this reaction endothermic or exothermic? Explain your answer. *(2)*

(c) For the reverse reaction write down the energy
   (i) used in bond breaking. *(1)*
   (ii) released as bonds form. *(1)*

This graph shows the energy changes which take place during a chemical reaction.

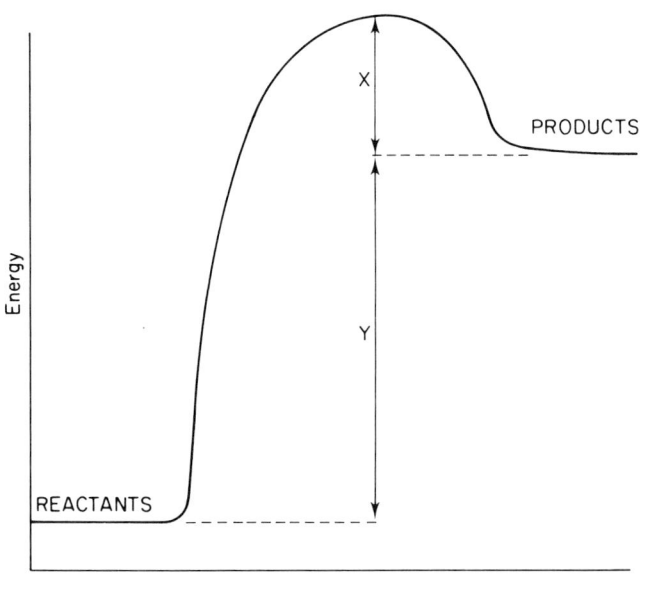

*4

Ammonia gas is produced by the reaction:

$$N_2(g) + 3H_2(g) \rightleftharpoons 2NH_3(g)$$

(a) What does the $\rightleftharpoons$ sign mean? *(1)*

(b) The Haber process uses this reaction to make large amounts of ammonia. The graph shows how the yield of ammonia in this process varies as the pressure used in the industrial process is changed.

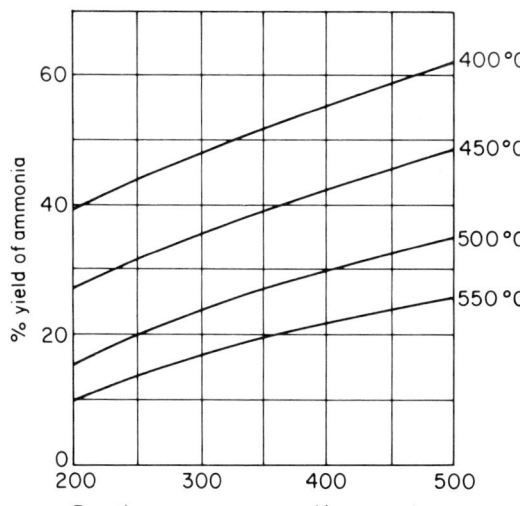

(i)   What does 'yield' mean? *(1)*
(ii) Describe the pattern linking reaction pressure to yield of ammonia. *(2)*
(iii) Give one advantage and one disadvantage of using a high reaction pressure. *(4)*

45

(c) (i) What pattern links the reaction temperature to the yield of ammonia? *(2)*

(ii) Give one advantage and one disadvantage of using a low reaction temperature. *(4)*

(d) In a newspaper story about developing countries, the writer said: 'Building an industrial plant to make ammonia is one of the most important things we can do to help.'

Imagine that you are writing a letter to the newspaper. Write to say whether or not you agree with the writer. Carefully explain the reasons for your view. *(10)*

Diagram 1 shows a bottle of mineral water. Use the information on the label to help you answer the questions which follow.

Diagram 1

St David's

BEACONS

Pure

NATURAL MINERAL WATER

BEACONS PARK

St David's spring is in the Beacons Park – over four hundred square miles of outstanding natural beauty, free from industry and intensive farming. Natural mineral water filters through the rock strata into aquifers deep beneath the mountains. It issues from our spring in all its natural purity. Low in minerals – only 82 mg per litre of dry residue – sodium and nitrates. Natural mineral water is a natural part of a healthy diet for you and your family.

Typical analysis	mg/l		mg/l
Calcium(Ca)	22.2	Sulphates(SO_4)	10
Magnesium(Mg)	3.2	Nitrates(NO_3)	1.3
Sodium(Na)	5	Hydrocarbonates(HCO_3)	67
Potassium(K)	0.4	Silica(SiO_2)	5.5
Chloride(Cl)	12	Dry residue at 80°C	82

A pure natural mineral water,
with a consistently low salt content

2 litres *e*

(a) The water in the bottle comes from an area which is 'free from industry and intensive farming'. Explain carefully how the presence of
(i) industry and
(ii) intensive farming
could have affected the water. *(4)*

(b) (i) Calculate the total mass of solids dissolved in this bottle of water. *(3)*

(ii) What mass of dry residue will be left if the water in the bottle is evaporated at 80 °C? *(1)*

(iii) Compare your answers to (i) and (ii) and explain the relationship between them. *(2)*

(c) Calculate the maximum mass of sodium chloride which could be in 1 litre of this water. (Relative atomic masses: Na = 23; Cl = 35.5) *(4)*

(d) Diagram 2 shows the solubility curves of some of the compounds found in the water.

Diagram 2

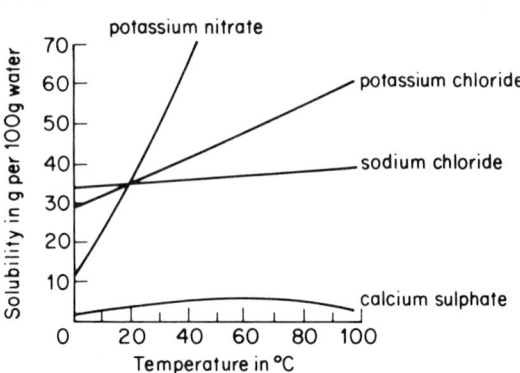

Use the curves in diagram 2 and your answer to (c) to comment on the claim on the bottle that it contains 'A pure natural mineral water with a consistently low salt content'. *(5)*

Pure gold is too soft to be used on its own. To make it useful it is alloyed with silver and copper.

'Carat' number		Percentage	
	Gold	Silver	Copper
22	92.0	5.2	2.8
18	75.0	12.5	12.5
14	58.3	20.85	20.85
9	37.5	31.25	31.25

(a) The percentage of gold in alloy of a stated 'carat' must meet a legal standard. What is the relationship between 'carat' number and the percentage of gold? *(2)*

(b) What percentage of '24 carat gold' would you expect to be gold? *(1)*

(c) English gold coins were made of gold alloyed with a little copper. Australian gold coins were made of gold alloyed with a little silver.
   (i) Suggest how these coins may have differed in appearance. *(2)*
   (ii) Some 9 carat gold contains 62.5% copper and no silver. Suggest one advantage and one disadvantage of this mixture compared with the 9 carat gold listed in the table. *(2)*

(d) Gold extracted from gold mines is usually alloyed with silver and sometimes with copper. At one time the silver was dissolved by using sulphuric acid on the alloy. This did not work if the alloy contained more than 33% gold. Such alloys had to be melted down with silver until the composition was about 25% gold. This was called quartation. Suggest why gold of less than 9 carats is not used. *(2)*

7 Liver salts

J-J's Liver Salts are a remedy for some minor ailments. They can be bought over the counter in a chemist's shop or at the local supermarket. The diagram shows the front label on a tin of J-J's Liver Salts.

On one side, the following information is given:
INGREDIENTS IN 100 g
sugar 40.5 g
with active ingredients
sodium hydrogencarbonate 22.6 g
citric acid 19.5 g
magnesium sulphate 17.4 g

(a) (i) Does J-J's Liver Salts contain any ingredients not mentioned on the label? Give evidence for your answer. *(2)*
   (ii) Explain why the sugar is not called an active ingredient. *(2)*

On another side of the tin the directions for use are given.

(b) (i) *'As a sparkling refreshing drink*: take one teaspoonful in a glass of water.'
What would you notice when one teaspoonful of J-J's Liver Salts are added to a glass of water? Briefly explain the chemistry involved in your observation. *(4)*
   (ii) *'For relief of upset stomach, indigestion and biliousness*: take one or two teaspoonfuls in a glass of water.'
How could J-J's Liver Salts relieve these disorders? *(2)*
   (iii) *'As a laxative*: two teaspoonfuls in a glass of water to be taken before breakfast or at bedtime. If a laxative dose is needed every day consult your doctor.'
Give a possible reason why it is taken before breakfast or at bedtime. *(2)*
Comment on the statement, 'if a laxative dose is needed every day consult your doctor'. *(2)*

(c) Is it possible to prove scientifically that J-J's Liver Salts 'wakes up your natural sparkle'? Explain your answer. *(3)*

*8 Flammability of textiles

Some types of cloth burn more easily than others:
Natural fibres
Cellulose fibres (e.g. cotton and rayon) burn easily. Protein fibres (e.g. wool and silk) do not burn easily.
Synthetic fibres (e.g. nylon and terylene)
These melt at between 150 °C and 300 °C which is below the temperature at which they burn. As they start to melt, they shrink. Once they have melted they burn if the temperature is high enough.

(a) (i) Cellulose is made of carbon, hydrogen and oxygen. Explain why it burns easily. *(3)*

(ii) A thin cotton shirt burns much more easily than thick cotton jeans. Burning often stops at a thick seam. Explain this. *(3)*

(iii) A coating of antimony oxide is used to 'fireproof' cotton. Antimony oxide melts at 620 °C and boils at 1150 °C. Suggest why it makes cottons 'fireproof'. *(2)*

(b) (i) Proteins contain nitrogen. Explain why wool does not burn easily. *(2)*

(ii) People in Bolivia live at high altitude. Foreign visitors feel very tired at this altitude because the air is so 'thin'. The Bolivians have more red blood cells and are able to cope. Body burns are rare in Bolivia because the woollen clothes people wear do not burn at all at this altitude. In fact wool does not burn at an altitude of more than 1000 m. Why is this? *(1)*

(c) (i) Hot ash from a cigarette falls onto a nylon shirt. What is likely to happen? *(2)*

(ii) A woman wears a nylon slip under a thin cotton dress. What is likely to happen if the dress catches fire? *(3)*

(d) Boys' clothing catches fire 2.3 times more often than girls' clothing. Suggest an explanation. *(2)*

*9 Magnesite

Kim Leung performed an experiment to find out how fast the mineral magnesite (magnesium carbonate) reacted with dilute nitric acid. She reacted 0.3 g of small pieces with 50 cm³ of dilute nitric acid and every two minutes measured the total volume of gas in the syringe.

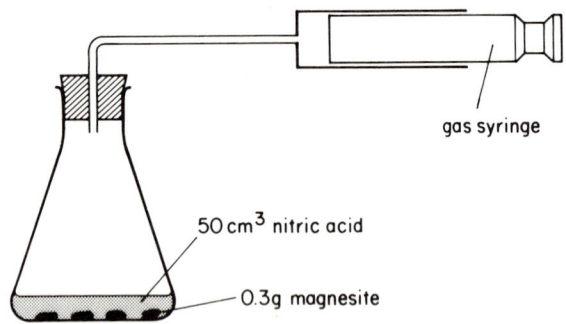

gas syringe

50 cm³ nitric acid

0.3g magnesite

The results of Kim's experiment are given below.

Time (minutes)	0	2	4	6	8	10	12
Volume of gas (cm³)	0	36	60	64	84	86	86

(a) (i) Plot Kim's results on a graph. Put time on the horizontal axis. *(4)*

(ii) Which reading did Kim get wrong? Suggest what it should be. *(2)*

(b) For the reaction write the
(i) word equation. *(1)*
(ii) formula equation. *(1)*

(c) Use Kim's results to calculate the volume of gas produced if 1 mole (formula mass in grams) of magnesium carbonate had been used. *(5)* (Relative atomic masses: H = 1; C = 12; N = 14; O = 16; Mg = 24)

(d) Magnesite is used as a medicine to treat dyspepsia (indigestion). People get dyspepsia when their stomachs produce too much acid.

(i) Magnesite can be given in tablet form or as a suspension of the powder in water. Which form would act the fastest? Explain your answer. *(3)*

(ii) Suggest why you shouldn't take more than the recommended dose of medicines used to treat dyspepsia. *(3)*

10 Crisps

Below is the front of the packet of a bag of Jim-Joy's potato crisps.

Jim-Joy's crisps

LOWER FAT POTATO CRISPS

LIGHTLY SALTED

65 gram

30% less fat

NO artificial preservatives colours or flavouring
REDUCED Salt   HIGH Fibre

On the reverse side of the bag the following information is given.

Lower Fat Crisps
'Jim-Joy's Lightly Salted Lower Fat Crisps contain 30% less fat and 25% less salt than standard crisps. They contain no colourings, no flavour enhancers and no preservatives. Less fat means fewer calories, though these crisps can help slimming or weight control only as part of an energy (calorie) controlled diet.'

The following nutrition information is also given:

Nutrition	Typical values	
	per 100g	per 65g bag
Energy	485k/CALORIES (2030k/JOULES)	315k/CALORIES (1320k/JOULES)
Protein	6.1 g	3.9 g
Carbohydrate available	60.4 g	39.3 g
Fat		
Polyunsaturates	5.0 g	3.3 g
Polysaturates	10.5 g	6.8 g
Dietary fibre	13.5 g	8.8 g
Added sugar	2.1 g	1.4 g
Added salt	0.9 g	0.6 g

Ingredients: Potatoes, vegetable oil (including hydrogenated vegetable oil), lactose, salt.

(a) Which unit is incorrectly written in the typical values data? *(1)*

(b) (i) What is the total mass of the contents of a 65g bag of Jim-Joy's crisps using the typical values data? *(2)*

   (ii) Suggest what else might be present. *(1)*

(c) Design an experiment to check the energy values given on the bag. *(4)*

(d) Design experiments to show that the crisps contain
   (i) 30% less fat than ordinary crisps.
   (ii) 25% less salt than ordinary crisps. *(8)*

---

11 Blood out of stones

Diagram 1 shows crystals of four pure substances.

Diagram 1

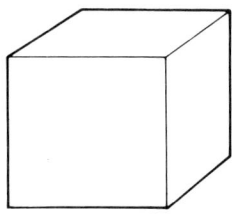

sodium chlorate

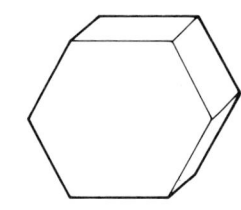

basalt

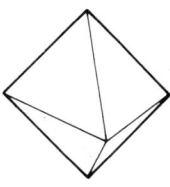

alum

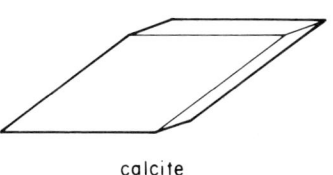

calcite

(a) (i) When a substance crystallises, the crystals must fit together to make a solid. Diagram 2 shows how this happens with basalt.

Diagram 2

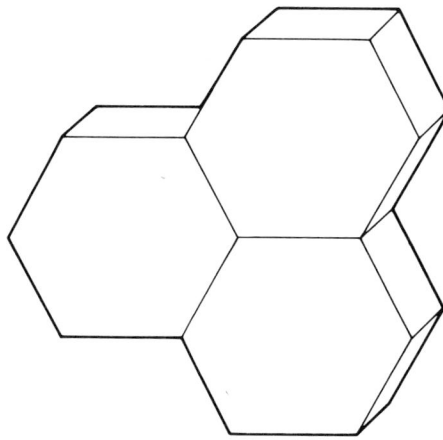

   Draw a diagram to show how calcite crystals fit together. *(2)*

   (ii) Can any crystals have the shape shown in diagram 3?

Diagram 3

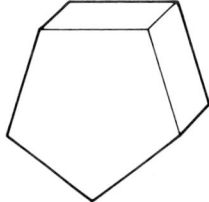

   Draw a diagram to explain your answer. *(3)*

(b) Impurities alter crystal shape slightly. A little borax was added to one of the substances whose crystal is shown in diagram 1. The substance was crystallised and diagram 4 shows the shape of the crystals formed.

Diagram 4

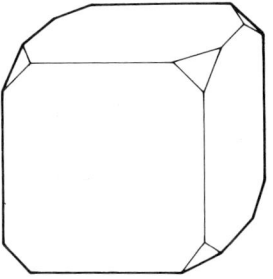

   Which substance do you think this was? *(1)*

(c) Scientists have examined traces of blood on stone axes used 6000 years ago. They have dissolved the blood and recrystallised it. They say that some blood on some axes comes from grizzly bears whilst that on others comes from sea lions or humans.
How do they know the difference? *(2)*

(d) Diagram 5 shows balls of plasticine in a jar of wallpaper paste.

Diagram 5

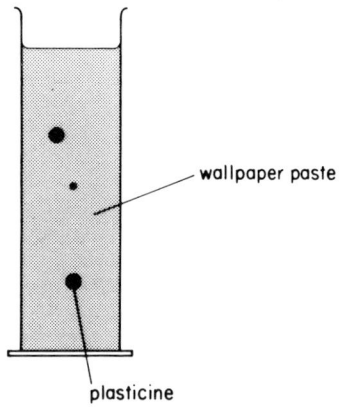

wallpaper paste

plasticine

(i) What force makes them fall? *(1)*
(ii) What force slows them down? *(1)*
(iii) Describe how you would find out whether big balls or small balls fall faster. *(4)*
(iv) Scientists have dissolved traces of blood from stone axes in a jelly. They have then used electrophoresis to find out what animal the blood came from. Positive and negative electrodes are put in the jelly and protein molecules move towards them. Suggest how scientists can use this to find out where the blood comes from. *(2)*

*12 Chlorofluorocarbons

Read the following passage and then answer the questions below.

Chlorofluorocarbons (CFCs) are compounds which have the same structure as alkanes except the hydrogen atoms have been replaced by chlorine and fluorine atoms. CFCs are more expensive than their 'parent' alkane to produce, so are only used where their special properties are required. CFCs are volatile, colourless, tasteless and non-toxic liquids. They are also non-flammable and unreactive. Diagram 1 shows the major uses of CFCs.

Diagram 1

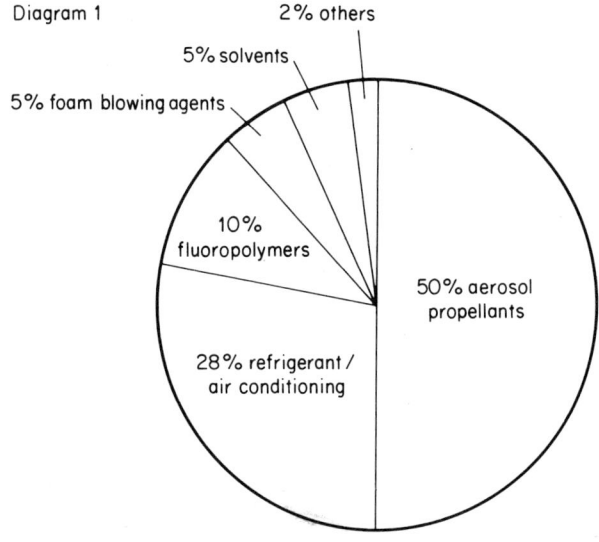

2% others
5% solvents
5% foam blowing agents
10% fluoropolymers
50% aerosol propellants
28% refrigerant / air conditioning

Most CFCs produced are used in cooling systems or to propel aerosols such as hair sprays. Some CFCs are used as foam blowing agents, e.g. to make expanded polystyrene containers for hot beefburgers.

Scientists believe that CFCs are damaging the ozone layer in the upper atmosphere. Many people who are concerned about the environment are refusing to buy aerosols containing CFCs. One fast food chain has stopped using expanded polystyrene for their beefburger containers.

(a) (i) What is an alkane? *(1)*
(ii) The structure of methane is:

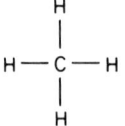

Draw the structure of the CFC dichlorodifluoromethane. *(2)*

(b) A company produces 1000 tonnes of CFCs per month. Draw a bar chart to show the amount of their CFCs used each month for each major use. *(4)*

(c) (i) What is a volatile liquid? *(1)*
(ii) Explain why CFCs are used in refrigerators and other cooling systems. *(3)*

(d) (i) Explain why CFCs are used in aerosols. *(2)*
(ii) The pressurised inhalers used by asthma sufferers contain a drug dissolved in CFCs under pressure. What properties must any replacement substance for CFCs have for use in inhalers? *(2)*

(e) Comment on the usefulness of the two measures, mentioned in the passage, which have been taken to reduce the amount of CFCs in the atmosphere. *(4)*

*13 Sodium carbonate the modern way

Sodium carbonate is a raw material for many processes. The world uses nearly 30 million tonnes every year. Much of this is manufactured from salt and limestone, which are cheap raw materials. The following equation summarises the reaction:

$$2NaCl + CaCO_3 \rightarrow CaCl_2 + Na_2CO_3 \quad [1]$$

However, calcium chloride and sodium carbonate will react together to give the reverse reaction. So the manufacturing process happens in several stages to avoid the reverse reaction. There are two main stages to the process.

Part 1

The limestone is heated in kilns to produce lime (calcium oxide) and carbon dioxide. The lime is mixed with water and the product is heated with ammonium chloride. This reaction produces ammonia gas. It is absorbed in brine (sodium chloride solution). The liquid formed is called ammoniacal brine. Ammoniacal brine is reacted with carbon dioxide:

$$NaCl + H_2O + NH_3 + CO_2 \rightarrow$$
$$\text{ammoniacal brine}$$

$$NaHCO_3 + NH_4Cl \quad [2]$$

The sodium hydrogencarbonate is precipitated and filtered out. The filtrate is recycled.

Part 2

The sodium hydrogencarbonate is heated to produce sodium carbonate, water and carbon dioxide. The carbon dioxide is recycled, the sodium carbonate is extracted and purified.

Sodium carbonate is manufactured in two grades 'A' and 'B'. Grade B sodium carbonate has a much lower density than grade A. It takes up a greater volume for the same mass. The tables show the major uses of the two grades of sodium carbonate.

Major uses of Grade A sodium carbonate		Major uses of Grade B sodium carbonate	
Flat glass, glass fibre	20%	Oils, fats, waxes, sugars	16%
Glass containers	63%	Heavy chemicals	57%
Sodium silicates	9%	Textiles	3%
Other chemicals	8%	Dyes, colours	3%
		Food, drinks	3%
		Other chemicals	18%

(a) (i) Write the equation for the reverse, competing reaction mentioned in the first paragraph. (1)
(ii) Explain carefully why this reaction 'competes' with the forward reaction. (2)

(b) Write the equation for the reaction which produces lime from limestone. (1)

(c) (i) Give the name of the chemical produced when lime is mixed with water. (1)
(ii) Write a balanced equation for the reaction between this chemical and ammonium chloride. (2)

(d) (i) What is contained in the filtrate from reaction [2]? (2)
(ii) Why can this filtrate be recycled? (1)

(e) Write the equation for the reaction which produces sodium carbonate from sodium hydrogencarbonate. (2)

(f) Draw two pie charts to show the uses of the two grades of sodium carbonate. (4)

(g) Suggest and explain an advantage of Grade A compared with Grade B sodium carbonate. (2)

(h) Draw a flow chart which summarises the process of manufacturing sodium carbonate from salt and limestone. (5)

14 Air in water

The table below gives information about gases: their solubility in water at different temperatures, their percentages by volume in atmospheric air and their relative molecular masses (RMM).

Table 1

Gas	Solubility (g in 100g of water)				% in air	RMM
	0 °C	20 °C	40 °C	60 °C		
Nitrogen	0.00294	0.00190	0.00139	0.001 05	78.9	28
Oxygen	0.00690	0.00434	0.00308	0.002 27	20.9	32
Carbon dioxide	0.335	0.169	0.097	0.0508	0.03	44

(a) The information shows that solubility increases with increase in
A. the temperature of the water.
B. the number of atoms per molecule.
C. the relative molecular mass.
D. the percentage of the gas in the air. (1)

(b) A can of a fizzy drink which has been stored in the refrigerator does not froth when the ring is pulled. A similar can which has been left in the sunlight produces a large amount of froth when opened.

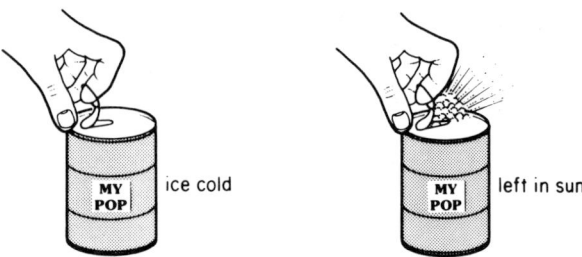

Explain these observations. (3)

(c) A sample of air dissolved in water at 20 °C was found to contain the following mixture of gases:

Table 2

Gas	Percentage (by volume)
Nitrogen	61.23
Oxygen	36.73
Carbon dioxide	2.04

The way in which a mixture of gases dissolves in a liquid can be calculated using Dalton's extension of Henry's Law.
(i) Use the information in the two tables to suggest what this law might say. (3)
(ii) Try to work out the data in table 2 from the data in table 1. (2)

A 15 Energy policy

Geologists think that there is enough oil left underground to last for only fifty more years.
(a) Assume that we continue to use oil at the present rate. Draw a graph to show how the quantity of oil remaining will alter over the next fifty years. (2)
(b) Some economists think that the price of oil will double every time the amount left underground is halved. Draw a graph to show how the price of oil might alter over the next fifty years. (2)
(c) As oil gets used up, coal and nuclear power will become more important. The USA and USSR have a lot of coal deposits. So has the UK. Very few other countries have big coal deposits.
   Some economists think that Britain should now build as many nuclear power stations as possible. Why could this be a good idea? (3)

(d) Some politicians believe that building nuclear power stations throughout the world would reduce the risk of nuclear war. Explain why this might be a reasonable idea. (3)
(e) Many environmentalists believe that electrical power should be produced from renewable sources.
   (i) What is a renewable source? Give three examples which could be used in Britain. (4)
   (ii) What arguments would an environmentalist use in convincing politicians to use renewable sources? (6)

*16 Sour beer

The triangular graph in diagram 1 shows the percentage (%) of the elements carbon, hydrogen and oxygen in four alcohols and their matching organic acids. Propanol contains 60% carbon, 13.3% hydrogen and 26.7% oxygen.

Diagram 1

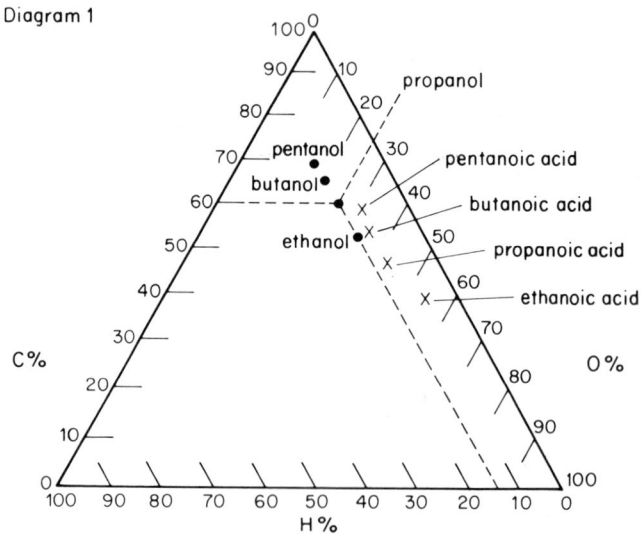

(a) (i) How does the percentage of the three elements vary in the homologous series of alcohols from ethanol to pentanol? (3)
   (ii) How does the percentage of the elements in the alcohols compare with that in their matching acids? (3)
(b) When a glass of beer is left to stand for a week, it goes sour. The ethanol in it changes to ethanoic acid. Malt vinegar is made by allowing enzymes in beer to catalyse the change from alcohol to acid.
   Where does the extra oxygen come from? (1)
(c) Ethanol is changed to ethanoic acid when it passes through orange crystals of potassium dichromate. A green salt of chromium is formed.

In a breathalyser the driver breathes through a tube containing crystals of potassium dichromate.

(i) Explain how the amount of alcohol in the breath could be found from the colour of the crystals. (3)

(ii) Dilute ethanoic acid is an electrolyte. It conducts electricity. A filter paper moistened with ethanoic acid also conducts electricity.

Some breathalysers have a battery and an ammeter to measure electric current. The meter gives a reading of alcohol level.

Draw a diagram to show how one of these breathalysers could work. Explain how it operates. (6)

*17 Relief

The Chemical Drug Company manufactures drugs. One of its products is RELIEF which is used to treat indigestion. RELIEF contains magnesium hydroxide.

The starting material for making $Mg(OH)_2$ is the mineral magnesite. Magnesite is magnesium carbonate with some solid impurities. This is the flow diagram for the process.

Diagram 1

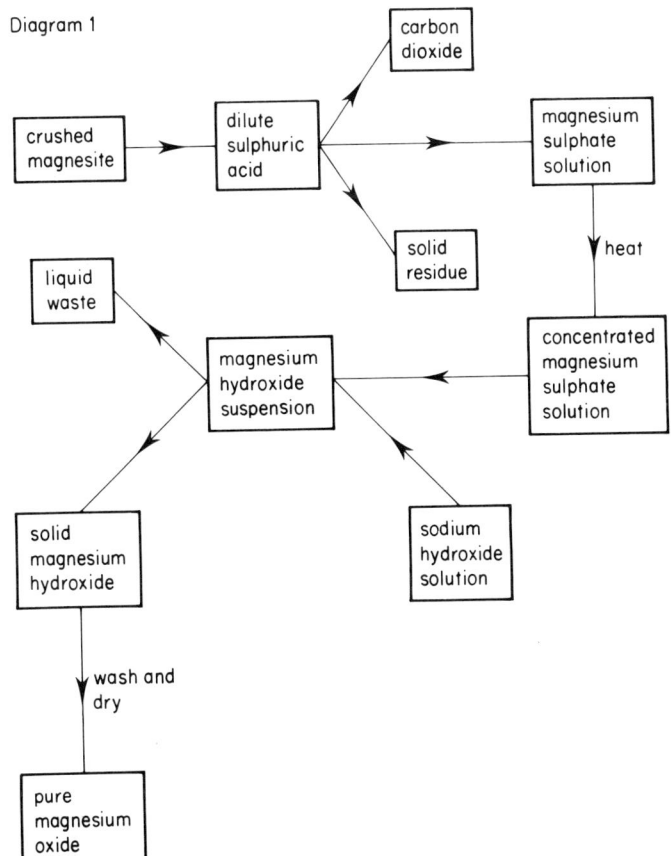

(a) The magnesite is crushed into very small pieces. Why are large pieces not used? (2)

(b) The reaction between magnesium carbonate and dilute sulphuric acid is given below:

magnesium + dilute sulphuric → magnesium + water + carbon
carbonate   acid   sulphate   dioxide

(i) Give two reasons why magnesite is added to the dilute sulphuric acid rather than adding sulphuric acid to the magnesite. (2)

(ii) After the reaction is complete the mixture is heated to its boiling point. Give a reason why this is done. (1)

(c) (i) The solid waste is separated from the liquid by:
A. chromatography.
B. distillation.
C. evaporation.
D. filtration. (1)

(ii) The process used to concentrate the solution of magnesium sulphate is called:
A. chromatography.
B. distillation.
C. evaporation.
D. filtration. (1)

(d) Before heating, the magnesium sulphate solution contains 10 kg of magnesium sulphate per 100 kg of water. A saturated solution at a temperature of 30 °C is needed.

Diagram 2

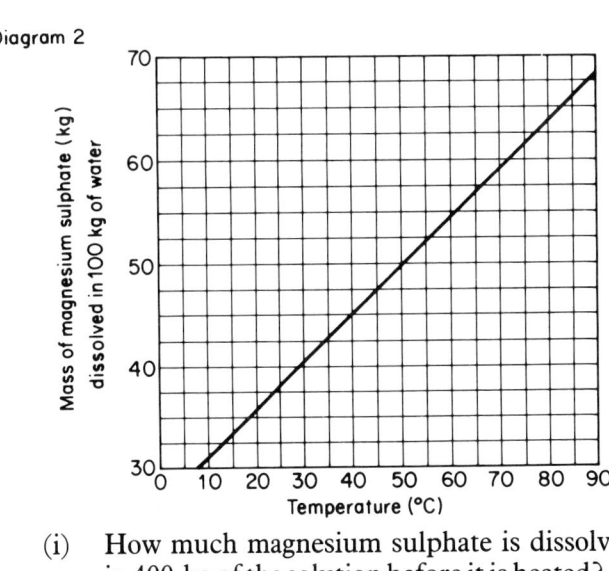

(i) How much magnesium sulphate is dissolved in 400 kg of the solution before it is heated? (1)

(ii) What mass of magnesium sulphate is dissolved in 100 kg of water when the solution is saturated at 30 °C? (1)

(iii) What mass of water has to be lost from the initial solution to give a saturated solution at 30 °C? (1)

(e) A concentrated solution of sodium hydroxide is added to the saturated magnesium sulphate solution. The following reaction occurs:

$$MgSO_4(aq) + 2NaOH(aq) \rightarrow Mg(OH)_2(s) + Na_2SO_4(aq)$$

Describe what you would expect to see when the sodium hydroxide solution is added. (2)

(f) After the magnesium hydroxide has been removed it is washed with water. Explain why. (1)

(g) Magnesium hydroxide decomposes at 350 °C to magnesium oxide and water.
  (i)   Write the formula equation for the decomposition reaction. (1)
  (ii) The magnesium hydroxide is dried by heating it to 200 °C. During the drying process the container containing the magnesium hydroxide keeps turning. Explain why this process is used. (2)

(h) Indigestion is caused by too much acid in the stomach.
  (i)   Explain why RELIEF can be used to treat indigestion. (2)
  (ii) Why could it be dangerous to take too much RELIEF? (2)

18   Mercury

Mercury is a poisonous heavy metal. It can cause damage to the brain and liver and, eventually, death.

(a) Here are some facts about the ways the human body absorbs mercury.
  1.   Mercury vapour can be absorbed through the lining of the lungs.
  2.   Most inorganic mercury compounds are almost insoluble in water. Those which do dissolve in water can easily be absorbed by the body through the gut wall.
  3.   All organic mercury compounds are soluble in fat and can be absorbed through the gut wall.

Use this information to answer the questions below.

  (i)   Ancient tribes used the dye vermilion (made from mercury(I) sulphide) to colour their bodies. Mercury(II) oxide is sometimes used in eye ointments. Comment on the solubility of these two compounds. (2)

  (ii) Metal tooth fillings contain about 50% mercury. The mixture has many remarkable properties. One is that it does not cause poisoning. Suggest two reasons why. (2)
  (iii) At one time people used hot solutions of mercury(II) nitrate to treat felt for making hats. Some of these people developed 'hatters' shakes' ('mad as hatters'). Suggest how they became poisoned by mercury. (1)
  (iv) About a hundred years ago people searching for gold found gold–mercury amalgam. They separated the gold by distillation in home-made stills. Some of them went mad. Suggest why. (1)
  (v)   Many people were poisoned at Minamata in Japan in 1953 by eating fish. At Minamata an inorganic mercury compound of very low solubility was released into the sea by a plastics factory. It is thought that bacteria, like those found in ponds, changed the mercury compound. Mercury was absorbed by shrimps. Fish ate the shrimps and people ate the fish.

      Explain what effect the bacteria must have had on the mercury compound. (2)

(b) Mussels and oysters concentrate mercury by a factor of 100 000.
The average concentration of mercury in sea water is 0.000 05 mg/kg. What would be its concentration in mussels and oysters? (1)

(c) Mr Jones absorbed 10 mg of inorganic mercury. The table shows how many milligrams of mercury were passed from his body each day by excretion.

Time (day)	1	2	3	4	5	6	7	8	9	10
Mass of inorganic mercury passed from body each day (mg)	1.09	0.96	0.87	0.78	0.67	0.61	0.55	0.48	0.44	0.39
Total mass of mercury passed from body (mg)	1.09	2.05								

  (i)   Copy and complete the table. (3)
  (ii) How much mercury remained in Mr Jones' body after ten days? (1)
  (iii) Draw a graph showing the total mass of mercury passed from his body against time. (4)
  (iv) Use your graph to estimate the 'half-life' of inorganic mercury in his body. ('Half-life' is the time it takes for the amount of poison in the body to be halved.) (2)

(d) Mr Halliday works with mercury and absorbs 2 mg per day. The graph below shows the mass of mercury which accumulated in Mr Halliday's body during a year.

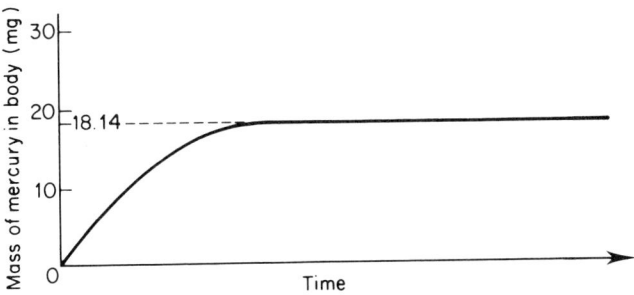

(i) Use the information in the table for Mr Jones to calculate the percentage of mercury which passes from Mr Halliday's body in the first day. (2)

(ii) Use this to explain the steady accumulation level shown in the graph (18.14 mg). (3)

(e) The 'half-life' of organic mercury in the human body is 70 days. For the same dose rate, the steady accumulation level is proportional to the 'half-life'. Estimate the steady accumulation level of organic mercury if the dose is 2 mg per day. (1)

Talking about the answers

1 Information handling

Information can be given in a table, a graph, a pie diagram, a bar chart, or a few paragraphs of writing. You may have to change information from one form to another: from a table to a graph or from a pictogram to a calculation.

You may have to collect together information from a number of different sources. For this you may have to use your knowledge of chemistry to decide what kind of information is relevant before selecting pieces of data.

Sometimes you have to do arithmetic, use formulae or work out percentages. You have to know what needs to be done and you must be able to do it correctly.

ANSWERS

2 In the pictogram, each bottle represents 200 thousand tonnes of sulphuric acid. You will see that some bottles are only half full. A half full bottle would represent only 100 thousand tonnes.
   (a) There are most bottles for the fertiliser industry. So the answer is **A**.
   (b) The pictogram shows one and a half bottles of acid for man-made fibres. The full bottle represents 200 thousand tonnes. The half bottle represents 100 thousand tonnes. So the total quantity of acid used is $200 + 100 = 300$ thousand tonnes. The answer is **C**.

4 Statement **A** is incorrect because copper has blue/green compounds and colours a flame green.
   Statement **B** is incorrect because calcium and sodium both have non-coloured compounds but do colour a flame.
   Statement **D** is incorrect. Flame colour is the same as compound colour for copper and zinc but not for calcium and sodium.
   Statement **C** is true. It correctly describes calcium and sodium. Notice that the statement says, 'compounds with no colour CAN colour a flame'. If it had said, 'they WILL colour a flame', it would have been wrong because zinc compounds do *not* colour a flame.

9 A molecule of Cu_3Sn contains three atoms of Cu and one atom of Sn. Its relative molecular mass is
$$(64 \times 3) + 118 = 192 + 118 = 310$$
   This is $\dfrac{192}{310} \times 100 = 62\%$ copper by weight. It is

therefore $100 - 62 = 38\%$ tin by weight. The ratio is
   62 parts copper : 38 parts tin.
Approximately 2 parts copper : 1 part tin.
Similarly, Cu_4Sn is
   68 parts copper : 32 parts tin.
Approximately 2 parts copper : 1 part tin.

The alloy which contains copper and tin in this ratio is speculum metal. There is more opportunity for the formation of Cu_3Sn and Cu_4Sn is this alloy than in the others.

Answer **C**.

25
(a)

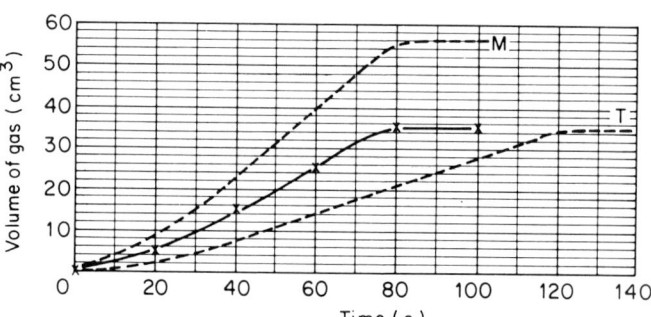

There is a mark for each axis, correctly labelled with a scale which allows all the points to be plotted. Two marks are for plotting the points. There are six points. Each error loses one mark. So five correct plots get one mark and four correct plots get 0. If you have chosen a scale for the axis which is difficult to use, (such as 5 small squares = 3 cm^3 of gas or 7 cm^3 of gas) you could lose marks here. How wrong does a plot have to be to lose a mark? Many examiners require points to be plotted correctly to the nearest half square of graph paper. If you plot the point as a large blob, it may not be possible to say where the 'centre' of the blob is to the nearest half square. If the plot is a tiny dot which disappears when the graph line is drawn, there may be no evidence of a point having been plotted. Points should always be plotted in pencil so that you can rub them out if they are wrong. A cross which is a vertical and a horizontal line can be difficult to see if the lines of the cross fit lines of the graph paper exactly. This is a particular problem on examination papers where the graph lines are grey rather than the brighter colours normally used for printing graph paper. Some people find that a diagonal cross shows a correct plot most easily. The final mark is for the graph line.

(b) (i) The graph rises to 35 cm³ and stays at that value. This is the total volume of gas produced in this reaction.

(ii) The graph stops rising when the time is 79 seconds. This is how long it takes for the reaction to finish.

(c) Hydrogen peroxide decomposes slowly at room temperature.

$$2H_2O_2 \rightarrow 2H_2O + O_2$$

The gas produced in this experiment is oxygen. Manganese (IV) oxide acts as a catalyst. It speeds up the reaction. The manganese(IV) oxide is not used up. There is as much left at the end of the reaction as there was to start with. Raising the temperature increases the speed of the reaction. Lowering the temperature slows the reaction down.

(i) Suppose two pupils had done exactly the same experiment, they would each have got the results shown in the table. The total quantity of hydrogen peroxide would have been double that used by one pupil. The total volume of oxygen every 20 seconds would have been double that given in the table. After 71 seconds, the reaction would end with the total volume of oxygen 70 cm³. So if you double the quantity of hydrogen peroxide, you get twice as much oxygen but the reaction takes the same time. The volume of oxygen produced is proportional to the quantity of hydrogen peroxide used.
Graph always shows a greater volume of gas than the original. *(1)*
Graph same shape as the original. *(1)*
Reaction finishes at the same time as the original. *(1)*

(ii) With the same quantity of hydrogen peroxide, the volume of oxygen produced is still 35 cm³. But the the reaction is slower so it takes longer for the 35 cm³ to be produced.
Graph starts by showing smaller volume of gas than original. *(1)*
Graph same shape as the original. *(1)*
Graph reaches same maximum value as the original. *(1)*
Reaction finishes later than the original. *(1)*

26 Cyclic graphs are useful for showing changes which are repeated every day, every year or in every cycle of any continuous process. For example, the changes of pressure in a cylinder of a four stroke internal combustion engine are repeated every cycle.

(a) The inner ring represents a rate of uptake of nitrogen of 1 kilogram per hectare per day. The next is 2 kg/(ha d), then 3 kg/(ha d).
  (i) 1. Grass needs most nitrogen at the end of May.
    2. Wheat needs most nitrogen at the end of May.
  (ii) 1. Grass first needs nitrogen in March.
    2. Wheat first needs nitrogen in March.

(iii) You should by this stage have found out how to read a cyclic polygon graph and can do this part of the question yourself.

(iv) 1. 'When grass starts to grow in spring, it grows faster and faster until it starts to flower at the end of May.' *(1)*
'The flowers are replaced by seed and the rate of growth is reduced.' *(1)*
'During September, which is a wet month, it starts to grow more quickly.' *(1)*
'As temperature falls and days shorten during October, it grows more and more slowly.' *(1)*
Any three of these statements would get the full three marks.
You need to understand that the word 'suggest' is telling you to put down some reasonable but not necessarily correct ideas. You are being asked for some good science guesses.
The suggestion:
'June, July and August are hot, dry months so grass does not grow as fast as in May and September,'
would also be reasonable and would get a mark. A mark scheme for a GCSE examination might simply say, 'Three reasonable and appropriate statements 1 × 3' and list the five given here as examples. However, during the marking of the examination, more than a dozen different suggestions may get credit.
2. What can you suggest about wheat?

(b) (i) You are told 'about 3000 kg' and 'about 98%' so strictly speaking you can get only an approximate answer. That is why the question tells you to estimate. But to get an answer you will have to do some arithmetic.
If 98% of nitrogen is tied up, 2% must be available to plants. *(1)*
2% of 3000 kg is
$$\frac{2}{100} \times 3000 = 60 \text{ kg} \qquad (1)$$

(ii) Ploughing will let air into the soil. *(1)*
Air is needed by the mineralising bacteria. *(1)*

(iii) 1. You should be able to answer this yourself.
2. And this.
3. 'Discuss' means that you have to put down some correct and relevant statements based on the information available. You need to link them to make a reasonable (but not necessarily correct) case for an argument or against the argument. In this case it doesn't matter whether you come out for or against the use of fertilisers. You may link evidence in such a way that you can say, 'There is no clear case for or against using fertilisers.'
A. 'Grass needs about 3 kg/(ha d) of nitrogen for half the year.' *(1)*

This is a correct conclusion from diagram 1(a).
A reasonable follow-up might be:

B. 'There are 365 days in a year so the nitrogen needed by grass in a year is

$3 \times \dfrac{365}{2} = 547$ kg/ha.' (1)

C. 'Two years of clover followed by one year of grass should give fairly good grass without any need for fertilisers.' (1)

D. 'Mineralisation gives about 1 kg/(ha d) for half the year.' (1)

E. 'So there is $1 \times \dfrac{365}{2} = 182.5$ kg/ha available by mineralisation.' (1)

F. 'About 40% of the nitrogen needed by grass can be produced by mineralisation.' (1)

G. 'Wheat needs less nitrogen than grass.' (1)

There are seven correct statements here based on the information given. But the authors feel that you should only get five of the six marks if they are presented like this.
The mark scheme might say:

'Four correct and relevant statements 1×4'
'Full coherence   2'
'(Partial coherence   1)'

Coherence means linking statements together so that it all makes sense. Statements A–B–C follow on from one another. D–E–F follow on from one another. But there is a break between A–B–C and D–E–F. And G is not related to anything else. So there is partial but not full coherence.

The marks in this question are for the skill of selecting relevant information and for the skill of communicating science ideas.

2 Problem solving

You may have to start by deciding exactly what your problem is. In question 11 a group of pupils are testing samples of coal to find which is the best. What makes a good coal: most heat produced by burning 1g, least smoke, least ash, easiest to ignite, cheapest? These pupils decided to investigate the amount of ash left. But the coal which leaves most ash may be cheapest. So is it a good coal or a bad coal?

Sometimes you have to find the cause of a problem. Your knowledge of Chemistry or a study of the data in the question may help you. Many of the questions in this section are marked with an asterisk ✱. You often need a knowledge of Chemistry basic principles to solve Chemistry problems.

You need to understand the reasons why a solution should work. There may be more than one possible solution and you need to be able to select the best from alternative solutions. One may be the quickest, another the cheapest and another the most effective.

There may be many occasions when your problem is an example of a general one. A set of formulae or rules may already exist. All you have to do is interpret the formula correctly for your particular problem.

5 (a) There was acid on the table. If the table had a surface which does not absorb liquids, the acid could be diluted and washed off. If the surface was wood, it could absorb the acid. This would make the wood deteriorate. So the problem is how to neutralise the acid. Flour (A) and salt (B) are neutral and will not affect the acidity of the liquid on the table. Vinegar (C) is an acid. If it is very dilute it could actually dilute battery acid but it will not neutralise it. Washing soda is alkaline. Sodium carbonate will neutralise sulphuric acid:

sulphuric + sodium → sodium + carbon + water
acid   carbonate   sulphate   dioxide

The correct answer is D.

(b) Battery acid is dilute sulphuric acid. But it is not very dilute. Neutralising it with an alkaline material can cause a safety problem.

A reaction will be less violent if the rate of reaction can be reduced. This can be done by reducing the temperature or reducing the concentration of the chemicals.

The problem occurs when the first bit of washing soda hits the strong acid. As this bit neutralises some acid, the remaining acid becomes less concentrated. So the reaction is less lively when the next bit of soda is dropped on the table. Adding less (A) or more (B) soda does not affect what happens when you drop the first bit.

Diluting the acid would certainly help by reducing the concentration. But if hot water (D) is used, the rise in temperature would cause a faster and more vigorous reaction than if cold water (C) is used.

The correct answer is C.

15 (a) Hundreds of years ago people realised that they would have a problem if they tried to burn limestone in a kiln which was also built of limestone. The inside of the kiln would burn away. (1)

Sandstone does not burn at the temperature used in the kiln. (1)

When the kiln at Jack Scout was built, sandstone would have had to be brought by horse and cart from about fifteen miles away and this would have made it expensive. But it was obviously cheaper to do this than to rebuild the kiln every few months.

(b) When the calcium oxide and silicon dioxide are heated to more than 1000 °C they combine to form slag. In the blast furnace the slag is molten and floats on top of the molten iron at the bottom of the furnace. It can be drained off to leave pure iron. If calcium oxide and sandstone reacted like this in a lime kiln, the sandstone lining would soon be destroyed. *(1)*
So temperatures in the Silverdale lime kiln must have been above 800 °C to burn the lime *(1)* but less than 1000 °C. *(1)*

19 (a) (i) Sometimes it is a good idea to be reminded of some Chemistry ideas before tackling a problem. An exothermic reaction is one in which heat energy is given out. *(1)*
An endothermic reaction is one in which heat energy is used up. *(1)*
(ii) The temperature in the tower is highest at a height of about 15 metres where the carbon dioxide reacts with the ammoniacal brine. *(1)*
So the reactions are exothermic. *(1)*
Temperature is lower at the bottom because of the cooling pipes.

(b) The Solvay Tower is a solution to a problem of industrial chemistry. Carbon dioxide must dissolve easily in the liquid which is in the tower. *(1)*
Sodium hydrogencarbonate must crystallise out from the liquid. *(1)*
Carbon dioxide is more soluble at low temperatures. *(1)*
Sodium hydrogencarbonate is less soluble at low temperatures. *(1)*

(c) Unused carbon dioxide *(1)* and water vapour *(1)* will leave the top of the tower as gases. There may also be some ammonia in the waste gases.

(d) (i) We use the word 'suspension' to describe solid particles held in a liquid. *(1)*
In this case, crystals of sodium hydrogencarbonate are carried in the waste liquids from the tower.
(ii) It must be possible to separate the crystals from the liquid. If they are too small they will pass through the filter, *(1)* and they are lost. *(1)*
(iii) 1. Small crystals are produced by rapid cooling *(1)* and you have to outline a way of cooling a solution rapidly. *(1)*
2. Larger crystals are produced by slow cooling. *(1)*
Again practical details are required. *(1)*

20 Smog is formed when chemical pollutants are trapped in the lower atmosphere. It is a serious problem. During the Great London Smog which lasted for 4½ days in December 1952, 4000 excess deaths were caused.
(a) This first part of the question involves understanding the problem by analysing data.
(i) The graphs for 7.00 a.m. and 3.00 a.m. show a simple temperature inversion: the highest temperature shown is no longer at ground level. The highest temperature is in fact at a height of about 80 metres. *(1)*
(ii) There is no temperature inversion at 7.00 p.m. By 11.00 p.m. a temperature inversion is developing. *(1)*
(iii) At 11.00 a.m. the temperature inversion is disappearing. *(1)*
(iv) The diagram must show the town between 11.00 p.m. and 11.00 a.m. *(1)*
(b) During the night the ground cools rapidly *(1)* if there are no clouds to prevent heat radiation escaping. *(1)*
(c) Tall chimneys allow smoke and waste gases to escape above the temperature inversion zone. *(1)*
(d) More smoke and pollutants would be produced during the night. *(1)*
They would become trapped by night-time temperature inversions. *(1)*
(e) (i) In valleys pollutant gases which cannot rise because of temperature inversion are also trapped by the valley sides. *(1)*
In flat areas, pollutant gases can spread over a wide area so they are less concentrated. *(1)*
(ii) Sulphur dioxide and other gases which cause respiratory illness are still produced, e.g. from car exhausts and industries. (One mark for the correct general idea, second mark for detail, e.g. which gases or what sources of pollution.)

3 Patterns

Most of the Chemistry Knowledge and Understanding needed for GCSE is linked together by patterns. The most important pattern is the Periodic Table. In 1817 Dobereiner noticed that groups of three elements like chlorine, bromine and iodine, or sulphur, selenium and tellurium have similar properties. He also noticed that the relative atomic mass of the middle one is the average of the relative atomic masses of the other two. Dobereiner called this the Law of Triads. In 1863 Newlands found that if elements are arranged in order of relative atomic mass, chemical properties are repeated in the eighth member of the series. He called this the Law of Octaves. These patterns are the basis of the groups and periods of the modern Periodic Table.

Sometimes information does not fit a pattern. It may be that the information is wrong. Newlands' ideas about octaves were not accepted at the time because relative atomic masses were not very accurate and that confused the pattern.

Sometimes information which does not quite fit a simple pattern can enable scientists to find a more accurate pattern.

In 1819 Dulong and Petit found that the relative atomic mass of an element multiplied by the heat, in joules, needed to raise the temperature of one gram of it by one degree Celsius usually equalled 26.46. Carbon, silicon and a number of elements did not fit this pattern. These deviations from Dulong and Petit's Law were explained by Einstein in 1907 as fitting the new Quantum Theory.

You can try out some of these patterns in question 17.

This is how knowledge of Chemistry has been built. Classifying information to form patterns, using them to make predictions, testing the predictions and getting better patterns is what Chemistry seems to have been doing for hundreds of years. Skill with patterns is the key to Chemistry.

3 (a) Non-metals Q and S both form negative ions. Q is like oxygen, it forms only the ion O^{2-}. So **D** is incorrect. S is like phosphorus, it forms a number of positive and negative ions. Phosphorus forms the ions $+3$, $+5$, -3 and several others as well. So **B** is incorrect.
Like many metals, P forms more than one ion. Iron forms Fe^{2+} and Fe^{3+} (iron(II) and iron(III)). **A** is correct: metals form only positive ions.

(b) Two elements E and G will form a compound EG if there are ions with matching charges E^{n+} and G^{n-}. PQ is possible (P^{2+}, Q^{2-}), RQ is possible (R^{2+}, Q^{2-}). PS is possible (P^{3+}, S^{3-}). PR is not possible. Answer **D**.

(c) A complete shell of electrons is very stable. The noble gases (helium, neon, etc.) have complete electron shells and until recently they were thought to form no compounds. Alkali metals have complete shells of electrons plus one electron in an outer shell. They very easily lose the outer electron to become Li^+, Na^+, K^+, etc. Alkaline earth metals have two electrons in their outer shell. These are easily lost to form ions like Mg^{2+} and Ca^{2+}. Element R seems to be of this type because it forms R^{2+} only. The answer is **C**. P is more complicated. Elements like iron have two electrons alone in an outer shell. These are lost to form an ion like Fe^{2+}. But the next shell is not actually full and one of the electrons in it is easy to lose. It can be lost to give an ion like Fe^{3+}.

12 (a) The rule or pattern you are using is that a more reactive metal displaces a less reactive metal from a solution of one of its salts. If you did not already know this you could work it out from the table so long as you know which is the more reactive of any two metals. For example, magnesium is more reactive than copper. You will see that magnesium metal added to a solution of copper ions does displace copper but when copper metal is added to a solution of magnesium ions, nothing happens.

So the most reactive of these metals is the one that is not displaced by any of the others. It is the one with a whole row of crosses: calcium. Similarly the least reactive has most ticks: copper. Arranging

the metals in order of increasing numbers of ticks gives the order of reactivity.

Solution containing ions of	Metal sample added				
	Calcium	Magnesium	Zinc	Iron	Copper
Calcium	×	×	×	×	×
Magnesium	√	×	×	×	×
Zinc	√	√	×	×	×
Iron	√	√	√	×	×
Copper	√	√	√	√	×

Marking: With lists like this, if you get two out of order, it is all wrong. So there are only two marks. A mark is lost for each metal which you get out of order.

(b) (i) X does not displace any, so it must be the least reactive. *(1)*
Y is displaced by all. So it must be less reactive than all of them. *(1)*

(ii) Potassium and sodium are more reactive than calcium. So X could be one of these *(1)* or another alkali metal.
Mercury, silver, platinum and gold are less reactive than copper. So Y could be one of these. *(1)*

(iii) The alkali metals are very reactive, *(1)* for example, they react vigorously with water, and handling them is very dangerous. *(1)*
The answer following Y will depend on the metal named in (ii). Silver, platinum and gold are precious metals *(1)* and are too costly to use. *(1)*
A correct property of mercury, e.g. mercury ions or vapour are poisonous *(1)* which causes a health hazard. *(1)*

(c) (i) Iron(II) chloride + zinc → zinc chloride + iron
or $FeCl_2$ + Zn → $ZnCl_2$ + Fe
or Fe^{2+} + Zn → Zn^{2+} + Fe

The question does not state what form of equation is required. Whichever you decide to use, you must be sure that you can complete it correctly.
Marking: products *(2)*
reagents *(1)*

(ii) Iron water tanks would tend to rust. *(1)*
The tanks are coated with a thin layer of zinc. *(1)*
This is called galvanising.
The zinc dissolves/reacts instead of the iron. *(1)*
(A zinc rod dropped into a tank made of bare iron would also have some effect.)

(d) Iron ships would rust, *(1)*
especially in salty sea water.
Magnesium stops the corrosion of iron. *(1)*
Explanation of why. *(1)*
There are a number of reasonable explanations. Following the earlier part of this question, it would be reasonable to talk about the reactivity series of metals

and magnesium ions being formed (i.e. it dissolves) in preference to iron ions (iron corroding). However, another explanation would call the magnesium a sacrificial anode, with iron/sea water/magnesium acting as a cell. As the magnesium dissolves to form ions, electrons flow to the ship through a metal chain or other conductor. The ship is the cathode and does not dissolve.

The two explanations are the same really. Some examination questions are deliberately open-ended and any reasonable explanation is accepted.

(e) Picks and shovels are iron.   *(1)*
     The orange coloured particles are copper.   *(1)*
     Explanation of displacement.   *(1)*

13 (a) (i) There is one mark for the obvious answer that energy increases as the number of carbon or hydrogen atoms increases. The second mark is for a more detailed description of the pattern.

Fuel	Energy given out (kJ/mol)	Increase in energy caused by an extra carbon atom (kJ/mol)
CH_4	890	–
C_2H_6	1560	670
C_3H_8	2220	660
C_4H_{10}	2877	657
	Total	1987
	Average	662.3

As the molecule increases by one carbon atom, the energy given out increases by about 662.3 kJ/mol.

(ii) 'Suggest a reason' asks you to make a reasonable guess. Marking schemes are usually very flexible. Correct and relevant science statements gain credit.

Each carbon atom produces one mole of CO_2, releasing the equivalent amount of energy.

Every two hydrogen atoms produce one mole of H_2O, releasing the equivalent amount of energy.

Two correct and relevant statements gain two marks. There are many other possible answers.

(b) (i) Hydrocarbons contain only carbon   *(1)*
         and hydrogen.   *(1)*
    (ii) These are alkanes.   *(1)*
    (iii)

methane   *(1)*
                                                     (1)

ethane   *(1)*
                                                     (1)

This part of the question is concerned with classifying the compounds. The alkanes have the general formula C_nH_{2n+2}. They form a homologous series. The formula increases by CH_2 on going from one to the next. Members of homologous series have similar properties which vary in a regular way through the series. In this question, heats of combustion increase in a regular way.

(c) (i) Working back down the homologous series, we get . . . . C_3H_8, C_2H_6, CH_4, H_2 subtracting CH_2 every time. We end with the hydrogen molecule as the lowest alkane! *(1)*

If the difference in heat of combustion is about 662.3 every time, the value for hydrogen should be $890 - 662.3 = 227.7$ kJ/mol.   *(1)*

The equation is:

$$2H_2 + O_2 \rightarrow 2H_2O + 531 \text{ kJ}$$

or

$$H_2 + \tfrac{1}{2}O_2 \rightarrow H_2O + 265.5 \text{ kJ}$$

The amount of energy produced depends on the conditions under which hydrogen is burned, but there seems to be some agreement between the value obtained from the alkane pattern and the 'proper' value.

(ii) 'Give a reason' means give any reason: it may not be the only one. Natural gas is readily available and cheap. It is mostly methane. Coal gas was mostly hydrogen, but it is no longer produced.

4 Evaluation

You will need to make judgements about statements by comparing them with data or with your knowledge of scientific principles. Some facts may have been overemphasised whilst others have been left out.

For hundreds of years people have believed that the hot spring waters at Bath (47°C) and Buxton (28°C) can improve health. The radioactive noble gas radon was discovered in 1900. It is produced deep underground by the radioactive decay of uranium and radium. It is very soluble in water and is contained in Bath and Buxton water. Radioactive material can be used in the treatment of some cancers. Some people believe that this gives a scientific basis for the medicinal properties of these waters. On the other hand, the hepatic waters at Harrogate contain dissolved hydrogen sulphide gas. Can you find out anything about how hydrogen sulphide affects health? You need to evaluate beliefs and statements in the light of scientific evidence.

Policy decisions in the chemical industry often involve an evaluation of estimates of costings and market prices in the future. Question 15 gives some costings for a plastic waste recycling plant which was considered to be worth building in Japan in 1971. In a different place, at a different time, the evaluation might be that such a plant could not make a profit.

Possible pollution effects of new technology have to be estimated by an evaluation of possible effects. Concorde (Question 20) could damage or improve the ozone layer. A number of chemical reactions are involved. Which ones win?

Solutions to problems can be judged. For example, why was a decision made to replace rubber–lead electric cables in household wiring with PVC–PVC cables? Question 18 gives you a chance to evaluate relevant data.

Practical procedures can be criticised. In Question 12 on wine making, have appropriate precautions been taken to ensure a wholesome product and to avoid the mess and danger of bottles exploding?

5   All four statements are true, but only one gives an explanation of why patients are not poisoned. When X-rays strike material, their energy is absorbed. There are several ways in which this energy can be absorbed. If an electron is knocked out of a target atom by an X-ray 'photon', this is called the photoelectric effect. Statement A tells us that the X-ray energy gets 'used up' by the barium sulphate. It explains why the negative photograph has a white patch where the barium sulphate was. But it tells us nothing about poisoning. $BaSO_4$ (Relative atomic masses: $Ba = 137$, $S = 32$; $O = 16$) has a relative molecular mass of 233. Most other salts of barium have smaller relative molecular masses ($BaCl_2 = 208$; $BaS = 169$) but a few do have a higher relative molecular mass ($Ba(NO_3)_2 = 261$). Barium absorbs X-rays well because it has a high relative atomic mass. But the relative molecular mass of its salts does not affect poisoning.

Anything which is undigested is passed out of the body in faeces a few hours after being swallowed. In fact all the barium sulphate will be passed out in this way as none of it is digested.

Barium sulphate is hardly soluble in water though it does dissolve slightly in sulphuric acid.

Salt	Solubility in water (g of salt in 100 g of water)
Barium iodide	220
Barium chloride	35.7
Barium oxide	3.9
Barium carbonate	0.002
Barium sulphate	0.0002

Things which dissolve in water in the intestines pass through the gut wall into the bloodstream. This is digestion. Because barium sulphate is not soluble, it cannot be digested so it cannot cause poisoning. The correct answer is **B**.

8   You are given some information about PVC and you have to evaluate it to decide how PVC will behave in different circumstances.
(a)   Window glass absorbs most of the ultraviolet rays.
                                                                (1)
So there will be no photodegradation in shops and kitchens.   (1)
In the street it is photodegraded by the ultraviolet rays in sunlight   (1)
(b)   Pigments (colourings) will absorb sunlight so the ultraviolet rays cannot penetrate into the PVC so well.   (1)
(c)   The surface layer absorbs ultraviolet rays.   (1)
This protects the rest of the PVC from damage by ultraviolet rays.   (1)

13   (a)   This involves value judgements. Sulphur dioxide and nitrogen oxides cause acid rain, which damages buildings and plants. These gases also cause respiratory illness. But carbon monoxide poisons by preventing the haemoglobin in blood from carrying oxygen, and some hydrocarbon gases cause cancer. So they are all pretty nasty.

However, sulphur dioxide comes from many other sources, including the sea. Reducing the 1% of sulphur dioxide in the air coming from exhausts to, say, ½% would not make very much difference to the total amount of sulphur dioxide in the air. But if the output of carbon monoxide, hydrocarbons and nitrogen oxides could be halved, there would be a significant drop in levels of these gases in the air.

Those are the three types of exhaust pollution which it is most important to control.

(b)

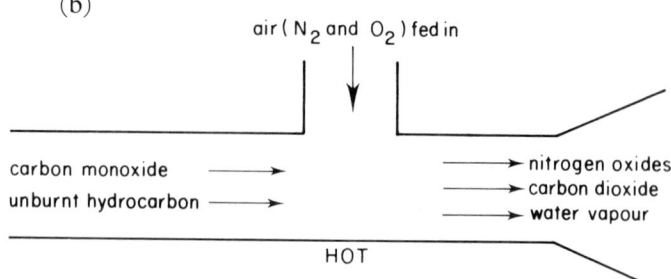

Unburnt hydrocarbon from the engine   (1)
will burn in the oxygen from the air which is fed in.
                                                                (1)

hydrocarbon + oxygen → water + carbon dioxide

Carbon monoxide from incomplete burning in the engine   (1)
will combine with oxygen from the air to form carbon dioxide.   (1)

carbon monoxide + oxygen → carbon dioxide

Nitrogen and oxygen in the air will combine at a high temperature to produce nitrogen oxides. *(1)*

nitrogen + oxygen → nitrogen + nitrogen
monoxide dioxide

(c) Better oxidation converts carbon monoxide to carbon dioxide *(1)*
and burns more hydrocarbon. *(1)*

(d) Fossil fuels contain some sulphur which burns to form sulphur dioxide. *(1)*
Alcohol contains only carbon, hydrogen and oxygen. *(1)*

(e) (i) The graph reading is between 0.5 and 1.0 ml/1000 l (or parts per million) of carbon monoxide.

(ii) Carbon monoxide is poisonous *(1)*
and its concentration would increase in a closed garage. *(1)*

(iii) The air/fuel ratio will be less than normal when the car is started. *(1)*
(This is ensured by using the choke control.) This gives a much higher concentration of carbon monoxide than normal. *(1)*

5 Synthesis

To tackle real situations in science you usually need to use several skills at the same time. In this section you may need to handle information, solve problems, find patterns, evaluate and apply knowledge of Chemistry all in the same question. Sometimes Biology, Physics or other areas of science are involved.

Question 11 is about identifying tiny traces of blood on ancient stone axes to decide which animals they were used on. This looks like a Biology problem. Or is it Stone Age History? It obviously relates to Forensic Science. In part (a) you are using information given about the shape of crystals and the way they link together to build up a pattern or identify shapes which will not form a pattern. Part (b) involves judgement about a crystal which does not fit a pattern perfectly. In part (c) you evaluate what you have found out about crystal patterns to suggest a way to solve the problem of identifying blood. Part (d) starts by presenting information about a Physics problem. You have to suggest what forces are involved: the everyday forces of gravity and fluid friction. Then you have to design a fair test to investigate the problem. In the final item you evaluate the Physics problem to find aspects which are relevant to analysing blood molecules which are moving under the influence of fluid friction and another force.

Multiple choice questions usually test one thing. So there are no questions in this section which are entirely multiple choice.

5 (a) To evaluate this statement you have to use appropriate Chemistry knowledge.
(i) Industry could cause pollution. *(1)*
Specific example: *(1)*
e.g. Tannery wastes contain chromium. Spoil heaps from coal mines produce sulphuric acid (because of oxidisation of sulphur in the coal).

(ii) Nitrates added as fertiliser or released by ploughing. *(1)*
Other pollutants, e.g. insecticides. *(1)*

(b) (i) This is an arithmetic problem.
There are two marks for adding together relevant figures given in 'typical analysis'. You need to handle the data correctly though. If 'dry residue' is included in your sum, you lose a mark. If HCO_3 is included, you lose a mark. Then you have to double the figure because the bottle contains 2 litres, not 1 litre. *(1)*

(59.6 mg in 1 litre, 119.2 mg in 2 litres)

(ii) The label says 'dry residue at 80 °C . . . 82 mg/l'. So if the 2 litres of water in the bottle are evaporated you should get 164 mg of dry residue.

(iii) The evaluation needs knowledge of Chemistry. The dry residue is greater than the mass of all the ions. *(1)*
The dry residue probably contains water of crystallisation. *(1)*

(c) This calls for arithmetic so that the problem can be dealt with using the familiar pattern of moles. Sodium ion is 5 mg (0.005 g) and chloride ion is 12 mg (0.012 g). But calcium chloride or sodium nitrate are also possible.
A mole of sodium chloride, NaCl, contains one mole of sodium ions and one mole of chloride ions.

We have $\frac{0.005}{23}$ = 0.000 217 mole of Na *(1)*

and $\frac{0.012}{35.5}$ = 0.000 338 mole of Cl *(1)*

So the maximum amount of sodium chloride is 0.000 217 moles. *(1)*
This is 0.000 217 x (23 + 35.5) = 0.0127 g of sodium chloride or 12.7 mg. *(1)*

(d) In putting together this answer you are using all the skills at once. You need to explain how you are interpreting salt: sodium chloride or what you get when you neutralise an acid with a base. *(1)*
You need to use the graph in some appropriate way *(1)*

e.g. sodium chloride solubility does not vary with expected variation in underground water temperature (perhaps 2–12 °C). Alternatively, only calcium sulphate and sodium chloride show constant solubility. This suggests some *consistency* of salt content. *(1)*

Comparison of quantities in a litre of Beacons water with the total possible (1)
e.g. sodium chloride in your answer to (c), 0.0127 g in 1 kg of Beacons water (1 litre) compared with a possible 360 g in 1 kg of water from the solubility graph. So *low* salt content (only $\frac{1}{28\,346}$ of possible sodium chloride levels). (1)

15 (a)

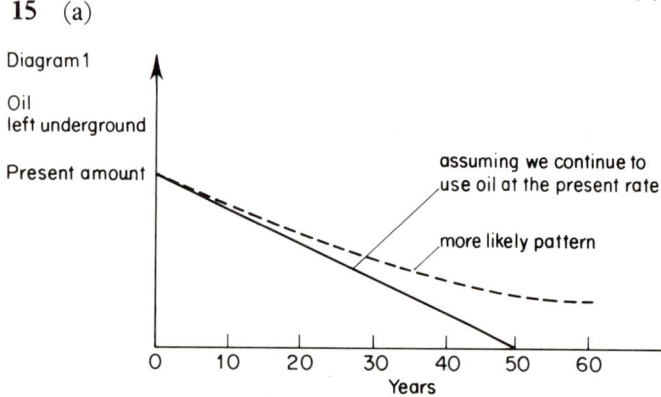

Diagram 1

You must convert the information given into a graph which shows an expected pattern. If we continue to use oil at the present rate, the amount left underground will decrease at a steady rate and there will be none left in fifty years.
Graph shows oil becomes 0 after fifty years. (1)
Graph is a straight line showing decreasing oil. (1)
Alternatives to oil will probably be used more and more, so a more likely pattern is that shown by the dotted line.

(b)

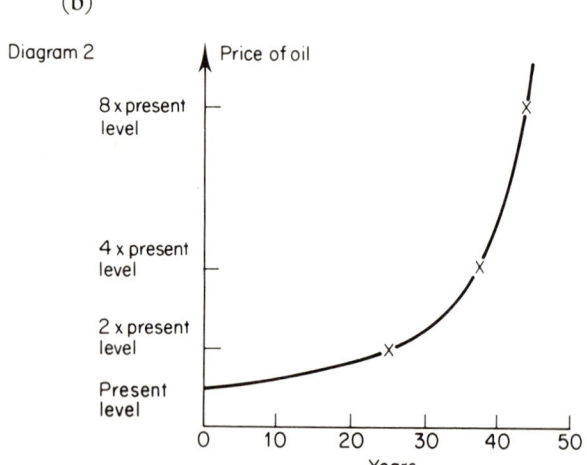

Diagram 2

The pattern is 'an exponential rise in oil prices'. You need to pull together information in your first graph and the idea of price doubling as quantity left halves. After 25 years there will be half the present amount of oil, so the price will be twice as high as at present (relative to wages, etc., at the time). After another 12.5 years (37.5 years from now) the quantity left would have halved again.

There will then be ¼ of what there is now. The price will be twice what it is after 25 years and four times what it is now. The oil remaining is halved again at 43.75 years, 46.875 years, 48.44 years etc. The price becomes 8,16,32, etc., times the present price.
Marking:
Graph shows doubling of price after 25 years. (1)
Completely correct pattern. (2)

(c) Planning for future profit requires accurate interpretation of data and patterns and the skill of evaluation. Oil can replace coal (as a source of organic chemicals and energy). (1)
So as oil gets used up, coal will become more expensive. (1)
Nuclear power stations would leave more coal available for sale at a huge profit. (1)
(There are enormous deposits of uranium relative to the amount of fuel needed by nuclear power stations. Fast breeder reactors convert unusable U-238 into the fuel plutonium.)
  The marking scheme for questions like this is usually flexible. Credit is given for reasonable and relevant statements linked together to form a clear argument.

(d) As oil runs out every country could be desperate to get control of the last remaining oil wells. (1)
There could be nuclear war between competing countries. Nuclear power would give energy without needing to use oil. (1)
Again, any reasonable statements which give a logical evaluation of this problem situation would gain credit.

(e) (i) Renewable sources do not get used up (1) because the real source of the energy will be there as long as people are on Earth.
From the sun:
  Solar energy. Heat can be absorbed to warm water, or light or heat can be used to produce electricity in photoelectric cells.
  Hydroelectric systems. The sun has evaporated water, mostly from the sea, which then falls as rain on high ground and fills up dams.
  Photosynthesis. Alcohol from sugar cane or sugar beet is used as a fuel for cars in some countries. Trees have been coppiced for hundreds of years. Timber was cut off them every few years so that they grew again for more timber to be cut off later. Fermenters produce heat and bio-gas (methane) as vegetable material rots down. The energy stored in plants comes from the sun.
Gravity:
Tides are caused by the gravitational attraction of water in the sea to the sun and moon.
Geothermal:
The centre of the Earth is very hot. Red hot

molten rock flows out in a volcanic eruption. In some areas water can be pumped into deep bore holes. It comes out hot.

Wind:

Air movement is mostly caused by two forces. There is a force from high to low pressure areas. The different pressure areas are caused by uneven heating by the sun. The other force is an inertial force on the air due to the Earth's rotation. Waves are caused by a disturbance of the surface of the sea or large lake. This is mostly due to wind.

Any three examples which could be used in Britain.   (*1 mark each*)
e.g. hydroelectric, wind and tides.

(ii) Five relevant science or other facts.   (5 x *1*)
Communication skill/logical argument   (*1*)
e.g. Use of reasonable sources would
      conserve oil, coal and other fossil fuels
                                              (*1*)

as continuing sources of organic chemicals.   (*1*)

International tension related to control of increasingly scarce fossil fuels would therefore be avoided.   (*1*)

Pollution resulting from the burning of fossil fuels would stop.   (*1*)

For example, there would be less production of sulphur dioxide so acid rain would be reduced.   (*1*)

This would reduce the cost of damage to health, crops and buildings.   (*1*)

The possible dangers of nuclear power could be avoided.   (*1*)

Britain could be self-sufficient and not need to import fossil fuels or uranium.
                                              (*1*)

There are eight relevant statements here. Five are enough to get full marks.

Acknowledgements

The authors are grateful to several authors and publishers for permission to use data which has appeared elsewhere. Sources are acknowledged in the text.

Thanks are also due to Julia Bennetts, who typed the manuscript of this book.

© 1990 J. Bennetts, M. Hannon & J. Mundie

First published in Great Britain 1990

British Library Cataloguing in Publication Data

Bennetts, J.
   Science skills : problems in GCSE chemistry.
   1.   Chemistry
   I.   Title II.   Hannon, M. III.   Mundie, J.
   540

ISBN 0 340 49080 2

Typeset in Plantin by Multiplex techniques ltd.
Printed and bound in Great Britain for Edward Arnold, the educational, academic and medical publishing division of Hodder and Stoughton Limited, Mill Road, Dunton Green, Sevenoaks, Kent by Butler & Tanner Ltd, Frome and London.